ROCK ROLES

Facts, Properties,
and Lore of Gemstones

Rock Roles

Facts, Properties, and Lore of Gemstones

Suzanne Bettonville

Lulu Enterprises, Inc.
Raleigh, NC

Rock Roles: Facts, Properties, and Lore of Gemstones

Copyright 2011 by Suzanne Bettonville

Cover design by Rio Dallapè
Cover photography by Suzanne Bettonville

All rights reserved. No part of this publication may be reproduced or transmitted in any form or by any means, electronic or mechanical, including photocopy, recordings, or any information storage and retrieval system, without written permission from the author. The author may be contacted at suzanne@ephemerala.com.

Visit the author's website at www.ephemerala.com.

Printed in the United States of America.

First printing: May 2011

ISBN: 978-1-257-03762-9

Published in 2011 by Lulu Enterprises, Inc., Raleigh, NC

*To Thomas Guensburg,
who made it all that much more interesting.*

table of contents

Introduction	ix
Part 1: Gemstone Facts, A-Z	1
Part 2: Additional Gemstone Information	113
Mohs Scale of Hardness	114
List of Traditional Birthstones	116
List of Ayurvedic Birthstones	117
List of Zodiac Stones	118
List of Anniversary Stones	119
Crystal Healing	120
Metaphysical and Healing Properties	122
Chakras	132
Chakra Stones	134
Index	136

Introduction

The history, lore, and healing properties of gemstones have aroused human interest for thousands of years. Gemstones have been valued for their beauty for as long as humans have known them, but many of them have played a part in mythology, religion, and medicine as well. Throughout history, humans have wanted to see value in gemstones beyond their use as ornamental objects. Thus, there has been amassed an enormous body of information about individual stones, by scientists, artists, medical practitioners, and hobbyists.

The casual gem collector or jewelry maker wanting to know a little more about certain stones may find himself or herself inundated with information from all angles. There are many books that deal specifically in one area of gemstone information, such as crystal healing or gem identification. Such books may contain adequate information for their particular field, yet may leave the reader wanting more information in other fields. Other books may contain a good variety of facts for each stone, but may be limited to only a few dozen of the most common stones, leaving the reader longing for information about some of the rarer gems.

This book provides comprehensive information covering topics such as healing properties of stones, locations where the stones are found, history of the stones, chemical properties, and instances where the stones appear in lore or mythology, for a huge variety of stones. Over one hundred stones are represented in this volume; each one with a single page (or in some cases, two pages) highlighting the stone's information in an easy to understand, concise format. In addition to gemstones, this book contains information on other materials that are often used in jewelry, such as precious metals, resins, and woods.

Crystal healers, jewelry makers, and inquisitive gem collectors can use this book as a quick informational guide. The stones are listed alphabetically by their most common names. The Index lists the common names plus many alternate names for stones, so anything that cannot be found in the main text will probably be found in the Index.

In addition to information about each stone, this book includes information on the Mohs scale of hardness, which is one of the easiest ways to identify many common stones, and which is mentioned many times throughout the book.

There are also lists of several types of birthstones, anniversary stones, and Zodiac stones. This is a good guide to turn to when in need of inspiration for a gift.

Also included is a special section for crystal healers and laypeople who need a quick guide on how to cure, heal, promote, or dispel certain physical or mental attributes. To use this section, look up the metaphysical, mental, spiritual, biological, inspirational, medical, physical, or emotional property for which you want help, and find the list of stones that are traditionally used for that purpose.

Finally, we have provided a brief overview on the concept known as chakras, found in traditional Indian medicine, including a list of stones found in this book that are useful for treating blocked or misaligned chakras.

The reader must note that the crystal healing section in this book is for informational purposes only. While gemstone therapy does no harm to the body, using it in place of proven medical practices can deny the body of medical help it needs. Please be responsible when considering alternate healing methods.

Minerals, rocks, and gemstones are fascinating objects that have inspired creativity and imagination for thousands of years. They open the observer to new worlds of art, culture, history, and science. They give the observer new ways to look at life and the world. Whether rough, polished, cut, cabbed, or carved, each stone is unmatched in its individuality and beauty. This book seeks to help the reader unlock the special roles of rocks.

gemstone facts

abalone

Abalone is a Mollusk (shelled fish) in the Gastropod class. Gastropods also include snails and whelks. Abalone is prized for its meat as well as its beautiful shell.

The pretty part of the shell is on the inside, where nacre (mother of pearl) of a variety of colors can be seen. The outside of the shell is dull brown.

The word "Abalone" comes from the Spanish word *aulone*, which is the name of the animal. The shell is also called "Paua Shell."

Crystal healers use Abalone for a large variety of purposes. It is thought to be helpful in disorders involving joints, bones, and muscles. It protects against uncooperative attitudes, stimulates creativity, and promotes calm. Not surprisingly, it also helps in calcium deficiencies.

Abalone and other shells are thought to be useful in facilitating decisions among groups of people.

Since Mollusks often have to leave their shells to find larger ones when they grow, all shells are considered to bring boundless growth in life and thought.

The Abalone that you see in jewelry is often a brilliant blue or purple. These specimens are dyed. The natural colors, while beautiful, are generally more varied, and may include blue, green, purple, pink, yellow, orange, and white. It is sometimes known as a "Sea Opal" due to the changing opalescent hues.

All jewelry made from shells, be it Abalone or Mother of Pearl, is suitable for Mother's Day gifts.

Abalone is revered in New Zealand, where it is called *taonga*, meaning "treasure." Commonly used for eyes in Maori carvings, the shells are thought to be the eyes of the ancestors looking down from the sky. The Maori tribe considers Abalone a good luck charm.

agate vs. jasper

Agates and Jaspers are both a type of microcrystalline Quartz called Chalcedony. (The Chalcedony group also contains Onyx and Carnelian.)

While many people believe that the color or pattern alone determines what the stone is called, the main difference is on a microscopic level. Agates contain microscopic *fibers* of crystalline Quartz, while Jaspers contain microscopic *grains* of crystalline Quartz. Thus, Agates are often transparent and Jaspers are usually opaque. The fibers in Agate are like the fibers in fiber optic bundles, transmitting light and making them transparent. The grains in Jasper serve to scatter the light, making them opaque.

Jasper can be thought of as "dirty Chalcedony," since it is dense with earthy or clayey matter. Jasper can be any color or many colors, based on the chemical impurities.

Agate most commonly occurs as the cavity fillings (nodules) inside geodes. It grows in consecutive layers, so Agates are usually striped. The bands are sometimes perfectly parallel, or they can be twisting and uneven.

Agates can be easily dyed to give them brilliant color, but Jaspers are so dense it is difficult to dye them.

Many rocks contain both Jasper and Agate. They are called "Jasp-Agates" or "Jagates." Crazy Lace Agate is technically a Jagate.

Most rocks sold today as Agates are actually Jaspers. People tend to prefer the sound of the word "Agate" to "Jasper," and jewelers know this, so they call many Jaspers Agates. When you have a stone with stripes, especially if they are transparent, it is safe to call it an Agate. If you have an "Agate" that completely lacks stripes or any sort of transparency, it is probably a Jasper.

Many of these stones have highly localized names, and the supply isn't always ongoing. This means that you may hear of a certain Jasper or Agate for a year or two, then it will fall off the market due to its depletion.

amazonite

Amazonite is a very rare variety of the most common mineral in the earth's crust, feldspar. It's the blue-green variety of the mineral Microcline. Amazonite is cherished for its lovely aqua color.

Almost always cut as cabochons, Amazonite sometimes displays schiller, which is a shimmering effect caused by reflections of inclusions within the stone.

Originally called "Amazonstone," named after the Amazon River in Brazil, the name was later changed to Amazonite. Incidentally, while it was named after the Amazon River, it is seldom found in that area.

Amazonite is sometimes confused with jade, a non-related stone. It was thought to have gotten its name from the story of Amazons (celebrated female warriors) giving the greenish stone to the men who came to see them. But the stone in question was probably jade.

Amazonite is a favorite of crystal healers. It is said to be a calming stone, reducing stress, balancing male and female energy, and promoting creativity. It is used to treat disorders of the nervous system, and resisting tooth decay and osteoporosis. It is said to bring joy to the wearer, and to make marriages happier.

Called "The Prosperity Stone," it is sometimes hung over doorways of businesses to bring in new business. It is worn by gamblers to attract money.

Found in many places over the world, the usual color for this stone is light sky blue, though the Amazonite found in Russia has a darker green tone to it. Amazonite that has whitish stripes cutting through the blue is called "Perthite."

While Amazonite is commonly used in jewelry, it is fairly soft and can easily chip and crack. It should be protected from scratches, pressure, and harsh chemicals.

Amber

Most gemstones are comprised of minerals or rocks (which are a combination of minerals). A mineral is described as a naturally occurring, inorganic substance with a definite chemical composition and crystalline structure. Amber, being organic in nature and without a specific chemical composition, is *not* a mineral. This stone bridges the gap between the plant and mineral kingdoms.

Amber is the hardened resin of the sap of prehistoric trees that were similar to today's pines and spruces. Many different types of trees created today's Amber (some evergreen and others deciduous), and therefore Amber can come in many colors and densities. The sap oozed out millions of years ago, and hardened into Copal. Over many years, the Copal went through some chemical changes to become the harder, more durable Amber. (Copal is often used for incense today, and it is softer and more easily dissolved in certain solvents.) Amber that contains bits of insects, plants, and feathers tend to be more highly valued, while the clearer Amber is less sought-after.

While we tend to think of Amber as yellow, orange, and brown, it comes in over 250 distinct colors, each with its own defining name. It can come in red, white, green, blue, black (a combination of other dark colors), and even purple. The transparent reds and greens are the most valuable, followed by the yellows and golds. Certain types of Amber fluoresce (glow under ultraviolet light), but these samples are hard to come by. Amber is a poor conductor of heat, and because of this, it always feels slightly warm to the touch.

Amber is a very soft gem (from 1-3 on the Mohs scale), so it is easily scratched. But it is also very brittle and will shatter if handled too roughly. It has the lowest specific gravity of any gemstone, and will float in salt water.

Amber is electromagnetic. When rubbed hard with a cloth, it will create static electricity that will pick up bits of paper.

Gem quality Amber is mainly found in two places: the region of the Baltic Sea, and the Dominican Republic. Differentiations between these two types of Amber are usually made by sellers.

The word "Amber" comes from the Arabic word *anbar*, which means "ambergris," which is not Amber at all, but a substance from whales which was used to make oils and perfumes. The Greek word for this gemstone was *elektron*, meaning "sun gold." The Greeks noticed that the stone had an electrical attraction, and their word for it would eventually be the base word for "electricity."

A lot of the Amber on the market is reconstituted. This means the natural Amber is melted down and re-formed into new stones. This makes it harder and more durable as jewelry, but it lacks the natural inclusions that make pure Amber so unique.

Amber is one of the oldest gems known. It was used for early medicinal purposes, and was thought to enhance the beauty of the wearer. Crystal healers use it to draw out negative energies, dispel depression, and promote physical, emotional, and spiritual health. It instills a feeling of self-confidence and a more positive mental state. It helps the wearer to find joy and humor, lightens the burdens of life, and promotes courage.

Amber appears in Greek mythology. Phaëthon, son of the sun god Helios, asked his father to allow him to drive the glowing chariot of the Sun one day. The searing heat of the chariot grew to be too much for him to handle, and the horses bolted, dropping him to earth, to land in a river and never be found again. Phaëthon's sisters were so distraught that Zeus took pity on them and changed them into trees. But they still wept, and their golden tears became Amber.

While fans of the book or movie *Jurassic Park* enjoy the idea of dinosaur DNA being found in the blood of trapped mosquitoes, extracting dinosaur DNA would be nearly impossible with the Amber specimens we have now. The Jurassic period was 144 million years ago, and dinosaurs died out completely 65 million years ago, while the oldest Amber specimens are only 50 million years old (and more often closer to 25 million years old).

amethyst

Amethyst is the purple form of Quartz. Traces of manganese create the purple color, while traces of iron are responsible for the depth of the color.

Amethyst can have different names, depending on the depth of the color. Light Amethyst is called "Rose de France," while dark purple Amethyst streaked with white Quartz is called "Dogtooth Amethyst."

This stone is purported to calm the mind, enhance meditation, relax the body, and enhance the dream state. It is a calming stone that soothes those in constant mental activity, and thus has earned the nickname of "Nature's Tranquilizer."

Amethyst has been thought to bring wealth to the wearer. In the Chinese philosophy of feng shui, Amethyst is put into the wealth corner to ensure the household of continued prosperity.

Physically, Amethyst is thought to prevent drunkenness, ease headache (if placed on the forehead) and diminish toothache. It also is thought to help cure insomnia when placed under the pillow. Ancient Greeks were so convinced it would allow a person to drink as much wine as they liked without getting drunk that they created wine goblets made from Amethyst.

Most Amethyst on the market today has been heat-treated to give it a deeper color. This treatment is permanent, and the color will not fade.

The word "Amethyst" is from the Greek *amethystos*, from *a* ("anti") and *methyein* ("to be drunk"), from *methy*, "wine."

In Greek myths, the god of wine, Dionysus, fell in love with a nymph, who ran from him. In a panic, she called upon Artemis, goddess of maidens, to save her from the drunken god. Artemis turned her into a white quartz statue. Later, Dionysus felt badly for what he had done. As a sign of affection, he poured wine over the statue. As the white turned to purple, he vowed that anyone who wore an amethyst would be immune to drunkenness.

Ametrine

Ametrine is a combination of Amethyst and Citrine in one crystal. It is created when iron in Quartz reaches two different states of oxidation within the crystalline structure. The exact cause of this phenomenon is currently not understood.

Most of the Ametrine on the world's market today comes from one mine—the Anahi Mine in Bolivia. There are rumors that the mine has recently run out of the gem and its future availability is uncertain, so Ametrine is projected to go up greatly in value in the next few years.

Ametrine is almost always cut in a faceted rectangular shape to highlight the differences in color. Recently, other cuts have been used, but the rectangle is still the favorite, and the best cut to display its beauty.

Ametrine was first known in the 17th century, when a Spanish conquistador received Ametrine as a dowry upon marrying a princess from the Ayoroeos tribe of Bolivia named Anahi. The conquistador brought the stone to Spain and gave it as a gift to the Queen of Spain, thus introducing it to Europe. The stone was virtually unseen by the public until 1980, when it came on the market out of Bolivia.

In addition to possessing the healing properties of both Amethyst and Citrine, crystal healers also use Ametrine to aid in removing toxins from the body, repairing DNA, and to help in the acceptance of physical change. It is thought to be helpful in releasing tension, stimulating creativity, achieving mental stability, and helping the wearer to remain calm under pressure.

The name Ametrine is obvious in its origin—it is a combination of Amethyst and Citrine. Prior to 1980, when it wasn't well known, its few rare specimens were called "Amethyst-Citrine Quartz," "Golden Amethyst," or "Trystine."

Ametrine, being Quartz, is very strong and stable; however, the colors can fade in sunlight.

apatite

Apatite is a fairly soft stone (5 on the Mohs scale) that comes in a variety of colors, though the most common colors are yellow and blue-green. A bright neon blue variety comes from Madagascar.

Apatite's softness, coupled with its tendency to be brittle, make it a rare choice for gemstones, as it can't be cut very well. Still, there are a few gem-quality stones. The darker the blue, the more valuable the stone.

This is thought of as a "Humanitarian Stone." It helps earth-conscious people to help Mother Earth, and it promotes good will toward and love for others.

The word "Apatite" comes from the Greek *apatos*, which means "to deceive." Since Apatite looks like several other stones, the ancient Greeks gave it this "deceptive" name.

Apatite can be transparent or opaque, so it can be cut into faceted stones or made into cabochons. Dark green faceted stones are called "Asparagus Stone," and dark blue faceted stones are called "Moroxite." When Apatite is cut into a cabochon that displays chatoyancy (cat's eye effect), it is called "Cat's Eye Apatite."

Blue Apatite is usually heat-treated to deepen its color. Green Apatite is left untreated.

Apatite used to be a source of phosphate for fertilizers.

Metaphysical healers use Apatite to reduce stress. It aids in alleviating "burnout," and quiets emotional upset during difficult times in one's life. It clears mental confusion, so it is a good stone to use while taking tests.

Crystal healers also find Apatite useful in decreasing hunger and promoting weight loss.

Aquamarine

Aquamarine is the blue-green form of the mineral Beryl. It is essentially the same stone as Emerald, but with a different color.

The word "Aquamarine" comes from the Latin for "sea water," due to its amazing ice-blue color. In ancient times, it was believed to be a good seafaring stone; sailors would wear Aquamarine to protect them during voyages. Legend has it that Aquamarine was treasured by mermaids, who would keep their lovely stones locked away until the time came to give them to sailors.

Crystal healers use Aquamarine to treat toothache, sore throat, sunburn, and fever (the latter two due to the stone's assumed "cooling" properties). It is said to help one to experience love and mercy, and to ease grief and depression.

Aquamarine is said to be helpful in preventing marital discord. It can help re-awaken love in long-married couples. Given to a spouse that is about to go on a journey, it ensures a safe return. It was also once thought to be a good gift to be given by a husband to a wife on the morning after their marriage.

Aquamarine's color is somewhat fragile; it tends to fade if left too long in the sun. It is sometimes heat-treated to deepen the blue color and enhance the clarity of the stone.

Due to its hexagonal crystal structure, Aquamarine is usually cut into rectangle shapes to bring out its natural brilliance.

It was once believed that to dream about Aquamarine was a sign that you were soon to meet new friends.

In the Middle Ages, Aquamarine was thought to be the best "oracle." Cut into a crystal ball, it was thought to be the optimal stone for foreseeing the future.

Aventurine

Aventurine is a form of Quartz that features tiny crystals, mainly mica, to give it a sparkling sheen. Different minerals determine the color of the stone. Aventurine comes in several colors, including green, red-orange, blue, and gray. Green is the most highly prized.

Aventurine is often color-enhanced to make its natural colors brighter. Lighter colors can also be dyed in shades not found in Aventurine naturally, such as purple or bright pink.

Aventurine got its name from a synthetic stone. This stone, which we call Goldstone today, was created accidentally when some copper dust fell into a vat of liquid glass. The stone was created *A venturi*, which is Italian for "by chance." Later, the name Aventurine was given to the *natural* stone that featured these sparkles, and the manufactured stone was given the name Goldstone.

Aventurine is sometimes called "Indian Jade," due to its resemblance to Jade and to the fact that much of it is found in India.

Aventurine is thought to be a lucky stone, especially for gamblers. It is said to attract money and wealth.

Considered a good all-purpose healing stone, Aventurine is said to promote calm and creativity, and to aid in healing eye problems such as nearsightedness. It is also helpful in problems with the heart, lungs, and muscles.

Aventurine was used to make tools such as axes by Ethiopian cultures some 2.5 million years ago.

Aventurine is used in jewelry and in carving. The German sculptor Helmut Wolf created a 5 ½ foot tall obelisk in Aventurine, which weighs 2 ½ tons.

Aventurine can fade if exposed to sunlight for long periods of time.

Azurite

Azurite is a copper carbonate that is a minor ore of copper. It is found all over the world in copper mines.

Azurite is closely related to Malachite. In fact, over time this stone absorbs water and actually *becomes* Malachite. Thus, Azurite and Malachite are often found in the same stone, the dark blue of Azurite contrasting with the deep green of Malachite. Occasionally the stone is also found with small flecks of red Cuprite in it, a variety known as "Bluebird."

Crystal healers use Azurite in applications of the head and brain. It is used to heal headaches, migraines, vertigo, and tinnitus. It's a good stone for removing toxins from the body.

Azurite is thought to bring the wearer an open mind, and the ability to let go of old destructive ways of thinking. It helps to promote change and make good choices. It inspires confidence and creativity, and helps the user retain information. For this reason it is a good stone for students to use.

The word "Azurite" comes from the Farsi (Persian) *lahzward*, which means "blue."

Ancient Greeks called Azurite "Caeruleum," and ground it into powder for use as medicine and dye. Europeans during the Middle Ages ground it for use in paint, and due to its close association with Malachite, some of the blue skies in these old paintings are starting to turn green.

Called the "Stone of Heaven," Azurite was sacred to ancient Egyptians and Native Americans. It was thought to promote communication with the spirit world. Egyptian priests would use ground Azurite to paint a third eye on their foreheads, which helped them communicate with the spirit world.

As beautiful as it is, Azurite is very soft, and is rarely used in jewelry without some stabilization from the addition of resins or waxes.

Bloodstone

Bloodstone is a dark green stone with spots of reddish iron oxide. It is sometimes called "Blood Jasper," although it isn't really a Jasper, but a different type of Chalcedony.

It is also called "Heliotrope." In ancient times, this stone when polished was said to reflect the sun.

Bloodstone is thought to enhance courage and promote understanding of the benefits of hard-earned victories, and therefore it is known as "Stone of the Warrior." It frees the wearer's energy to explore the unknown and take chances, and it is a good stone for those who are self-employed.

This is a very porous stone, and will reject its polish if placed in water.

In ancient times, Bloodstones were thought to slow bleeding. They were used to assist in labor and delivery, and to purify toxic blood.

Bloodstone is a calming stone, removing distractions that might cause a person to walk into dangerous situations.

Bloodstone has been believed throughout history to be an aphrodisiac, and even today it is ground into a fine powder for this use in India.

The blend of red and green, which are opposites on the color wheel, are thought to allow Bloodstone to help integrate and blend opposites. This makes it a good stone to give to lovers who are very different from each other.

Legend has it that Bloodstone was formed when drops of Christ's blood fell on green Jasper at the foot of the cross. Thus, it is a highly revered stone among certain Christian groups. Medieval Christians often used Bloodstone to carve scenes of the crucifixion and martyrs, which helped it gain the name "The Martyr's Stone."

Blue Lace Agate

Blue Lace Agate, also known as "Banded Chalcedony," is comprised of white and light blue bands that look dull in their natural state but are brought to beautiful life when polished.

The word "agate" is thought to come from the name of the Achates River (now known as the Drillo) in Sicily. The Blue Lace name refers to its delicate lacy look.

This stone was first discovered by a Namibian farmer named George Swanson, on his farm (known as Ysterputs Farm), who dubbed it the "Gem of Ecology" due to its white and light blue banding resembling the earth's white clouds over blue sky. The best Blue Lace Agate in the world still comes from Ysterputs Farm.

Blue Lace Agate is said to be a calming stone. It helps the wearer achieve peace and wisdom. It is a gentle stone, and when worn by gentle people, it helps them to not be changed by the hectic events surrounding them. A Blue Lace Agate in the house reduces household tensions.

Crystal healers use Blue Lace Agate in applications of the throat. It is said to soothe a sore throat, and reduce hoarseness of the voice, as well as curing problems related to speech such as stuttering. It is also said to be helpful in treating migraines.

In addition to aiding in throat problems, Blue Lace Agate is thought to be helpful in problems of the bones.

Being a throat stone, Blue Lace Agate is thought to be an aid in communication and is a good stone for public speakers to wear. It is called the "Stone of the Diplomat," and is thought to help the wearer find the best forms of self-expression through diffusing hostile situations and finding the best words to speak.

Blue Topaz

Blue Topaz is naturally a pale blue stone. The color is heightened by heat treatment, which sometimes occurs in nature but is usually done artificially. There are three basic colors of Blue Topaz: Sky Blue (the lightest), Swiss Blue, and London Blue (the darkest, which is often used as an inexpensive substitute for Sapphire). Pale Blue Topaz is also occasionally dyed blue for the most inexpensive gems. Truly blue stones are relatively rare. More often, white or pale yellow Topaz stones are heated to get them to turn blue.

Because of the ease of artificially coloring Blue Topaz, there is a lot of it on the market, in very precise colors that can be well-matched in jewelry. It makes a good, inexpensive alternative to Aquamarine.

Blue Topaz is seen as a cooling, calming stone. It was believed to have cooling properties, both physical (it was thought to cool boiling water), and emotional (it is said to calm a hot temper).

Blue Topaz is used by crystal healers to aid in communication. It helps to clear muddled thoughts, clarify the mind, and put ideas into words. It helps bring leadership qualities to the wearer.

Some Blue Topazes have been discovered to be radioactive. Because of this, the Nuclear Regulatory Commission requires all Blue Topaz mined in the United States to be tested for radiation levels. If you obtain a Blue Topaz that was mined somewhere other than the United States, it is a good idea to get it tested for radiation levels.

Blue Topaz was once thought to render the wearer invisible.

While Topaz is a very hard stone, it has perfect lines of cleavage, which means it can be broken by one swift blow. Care should be taken to protect Topaz stones from injury.

Blue Topaz is said to be a lucky stone for people born in the 4th hour of the morning.

Botswana Agate

Botswana Agate is a banded stone comprised of layers of Quartz-family components (crystal Quartz, Amethyst), Chalcedonies (such as Onyx and Carnelian), and various Jaspers.

Botswana Agate is also known as "Eye Agate" due to the bulls-eye pattern that can sometimes be found when it is cut just right. For this reason, it was thought to ward off the Evil Eye.

Known as a "change stone," Botswana Agate is thought to be helpful in coping with changes in life. It helps the wearer to see change in positive ways, and to find solutions to problems.

Botswana Agate is thought to help the wearer combat feelings of depression or guilt. It is a comforting stone that fosters honesty and trust. It can help the wearer to feel less hurt, and is a good stone to wear when you are feeling sensitive or lonely.

This stone is said to enhance creativity, help achieve success, aid in concentration, and improve memory.

Crystal healers use Botswana Agate to help rid the body of toxins, and to help heal broken bones.

Botswana Agate is thought to enhance and improve sexual energy by providing stamina. It is considered a very sensual stone due to the curving patterns of the layers.

The name comes from the country of Botswana, where the stone is mined.

In Botswana, it is worn by children to protect them and to help prevent them from tripping.

Brecciated/Poppy Jasper

Brecciated and Poppy Jaspers are rocks that have been broken into small pieces, and then "glued" back together by other rocks. They consist of angular particles that are larger than sand, surrounded by finer grained matrix.

The name "Brecciated" refers to the structure of the stone. Red Jasper is usually a component, but Brecciated Jasper can contain any of many different types of Jasper. The stone commonly called Brecciated Jasper has fairly large (6-15mm) chunks, whereas the stone commonly called Poppy Jasper usually has smaller (1-5mm) chunks.

The nature of the stone—where a formerly existing stone had been broken and fixed—is very significant to crystal healers, who use the stone to help heal any "broken" part of the body that has been "glued" back together, such as broken bones. Its heavy concentration of Red Jasper also makes it an excellent stone for blood disorders.

The name Breccia comes from the Old Teutonic *brekan*, which means "to break." Poppy Jasper is so named due to its resemblance to the red flower.

Brecciated and Poppy Jasper are thought to aid in dream recollection, provide a playful, joyous attitude, and to break any "drought" the body may be experiencing, whether it be creative, intellectual, fear, or acceptance. The stone produces an adrenaline-like awakening to areas of the body that are inactive. The stone is not recommended for use by pregnant women.

These Jaspers are thought to increase physical endurance and protect against dehydration.

This stone is thought to be very animal-friendly. It aids in communication with animals, and makes an excellent stone for people in the animal business, such as veterinarians and farmers. It is also an aid for animal-based allergies.

calcite

Calcite is one of the softest minerals, and one of the most common on earth. It is often used in cements and mortars, and is the major component in Limestone and Limestone's igneous cousin, Marble.

While most minerals show only a few crystal formations, Calcite comes in over 300 different crystal forms. It is also one of the most colorful stones, covering all areas of the rainbow, and some are even iridescent. Calcite fluoresces purple or bright red under a black light, and also displays the interesting property of triboluminescence. This can be seen when Calcite is struck—it will glow in a dark room.

The word "Calcite" comes from the Greek *chalix*, meaning "lime" (as in "limestone").

Crystal healers use Calcite as an aid in memory and to enhance learning abilities. It is said to increase prosperity and aid intuition.

Calcite has electrical properties. It shows electrical impulses under pressure, and is thus believed by crystal healers to amplify energy in the body.

The form of Calcite called "Optical Calcite" or "Iceland Spar," which is the clear variety with smooth crystals, shows double refraction. Light beams enter the crystal and split into a fast and a slow beam, which leave the crystal at different angles. Thus, when you lay a Calcite crystal against writing or a line on paper, a perfect double image will be seen. This double image is said to double the healing properties of the stone.

You can see some amazing samples of Calcite simply by visiting a cave. Most cave formations are the result of Calcite in the rock. Calcite is very soft and dissolves quickly, thus helping in the creation of stalactites, stalagmites, arches, and a host of other cave formations.

Calcite is extremely soft, which means great care must be taken to protect it when it is used in jewelry.

Carnelian

Carnelian (also spelled "Cornelian") is a form of Chalcedony (Quartz), and is sometimes known as "Sard" and "Pigeon Blood Agate." When banded with white stripes, it is called "Sardonyx." Its red color comes from iron in the Quartz.

Carnelian is said to protect against anger, envy, and fear; it is a calming stone. It enhances creativity, aids memory, and helps the wearer to stay grounded in the present and to make good judgments.

The energies of Carnelian influence the reproductive organs. Thus, it is commonly used as a symbol for fertility.

In ancient Egypt, Carnelian was associated with the heart, and the resurrection of mummies. Egyptians used it as a form of protection for both the living and the dead. They would put a Carnelian carving of the goddess Isis at the throat of a mummy to ensure a safe passage to the next life.

In the Middle Ages, it was used as protection against nightmares, mind-readers, and lightning.

The word "Carnelian" comes from the Latin *carne*, meaning "flesh," due to the reddish-brown color range it shows. The reddish color becomes more intense when exposed to sunlight.

Natural Carnelian is often dyed to help improve its color. A lot of the "Carnelian" on the market today is actually Chalcedony that has been dyed.

A dream in which you see Carnelian is said to be a warning of future problems.

Carnelian is sometimes called the "Actor's Stone." It is thought to help in the theatrical arts, and is a good stone for a performer to wear.

Chrome Diopside

Diopside is a mineral that is found in metamorphic limestone rocks, igneous basalts, and meteorites. It comes in a variety of colors, but its most notable gem-quality material is a rich green, colored with chromium, which is called "Chrome Diopside" or "Chromium Diopside."

While Chrome Diopside ranges in color from deep dark green to colorless, its preferred color is green. It has a deep color saturation, so the larger the stone, the darker the color. For this reason, you rarely see good gemstones that are larger than 5 carats, as the color in larger stones tends to look black rather than green. It is sometimes chatoyant (shows an "eye" due to layers of Rutile within it), and these stones are cut in cabochons, but most Chrome Diopside is faceted.

Chrome Diopside is mainly mined in Africa and Russia. It is sometimes referred to as "Russian Diopside" due to its location.

This gem is fairly soft and heat sensitive. It should be kept from sudden extreme temperature changes and chemicals. It is generally not suitable for everyday wear.

The name "Diopside" comes from the Greek *dis,* meaning "two," and *opsis,* meaning "vision." This is due to its occasional pleochroism (the property of looking a slightly different color when viewed from different angles).

Called the "Crying Stone," Chrome Diopside is thought to aid in healing traumas by bringing cleansing tears. It helps in mending relationships and improving the ability to trust. Wearing the stone helps to ease deep fears.

Crystal healers use Chrome Diopside as an aid in problems of the heart, lungs, and circulation. It is thought to improve intellect, analytical abilities, and creativity.

This stone is sometimes sold under the trademark name of "Vertelite," which is Latin for "green tone."

Chrysocolla

Chrysocolla is a copper-rich mineral that is closely related to Turquoise, and is sometimes confused with Turquoise. It can be found wherever copper deposits are located, which includes the southwestern Untied States. It occurs in crusts, masses, and tiny needle-like crystals.

The high copper content gives Chrysocolla a brilliant green-blue hue. As it is a hydrous stone, it dries out easily and is prone to breaking. It's a very soft stone (2-4 on the Mohs scale of hardness), but is often found in Quartz, which makes it hard enough to cut. Still, it needs to be protected when used in jewelry. Sometimes the stone itself can change with changes in humidity. Its color can become slightly different, and its opacity can change when the amount of water in the stone varies.

Crystal healers use Chrysocolla to aid physical strength and endurance, and to ease allergies. It is used in applications of the bone, and to heal arthritis, ulcers, and digestive problems.

Chrysocolla is thought to help relieve stress and grief, to build up broken relationships, to heal broken hearts, and to expel guilt. It is thought to help the wearer gain wisdom in speech—to know what to say, when to say it, and when to stay quiet.

The name "Chrysocolla" comes from the Greek *chrysos*, which means "gold," and *kolla*, which means "glue." Ancient Greeks used certain stones as solders for metals (including gold), and copper-rich minerals were among these.

In the 1950s, Chrysocolla was voted by lapidarists (gem cutters) as the Most Popular Gem.

The ancient Egyptians called Chrysocolla the "Wise Stone." Cleopatra was thought to have carried it with her at all times.

Chrysocolla is sometimes sold under the name of "Gem Silica."

Chrysoprase

Chrysoprase is the most valued of the stones in the chalcedony group. The stone has a bright apple green color due to a high nickel content.

The most valuable specimens come from Australia, and thus Chrysoprase is sometimes called "Australian Jade."

The name "Chrysoprase" comes from the Greek words *chrysos* ("gold") and *prasos* ("green").

Chrysoprase is a wonderfully durable stone that resists chipping and cracking, so it can be worn everyday. The bright color can fade if exposed to heat and sunlight, but can be restored with moist heat.

Crystal healers use Chrysoprase for problems relating to the reproductive system. It is also thought to help heal any type of wound. The stone should never be placed directly on the wound, but should be held above it.

Chrysoprase is a calming stone that helps the wearer work through problems without the distractions of daily life. Thus, it is a good stone for children to wear in school. It also helps to tap into inner courage, to gain self-confidence and create better social interaction. Its help in clarifying problems and providing solutions make it an ideal stone for those in positions of management.

Chrysoprase is thought to shield the bearer from negative energies, and placing a bowl of Chrysoprase by the entrance of the home helps to keep the home a happy one.

In the 1800s, thieves believed that holding Chrysoprase in their mouths would render them invisible.

Romanian folklore says that Chrysoprase enables the wearer to understand the language of lizards.

chrysotile

This is a stone that is almost *never* used for jewelry. The Chrysotile beads that are occasionally found on the market are really pretty, so perhaps Chrysotile as jewelry will eventually catch on.

Chrysotile is the main component in asbestos. There are two types of asbestos. One has hard, straight fibers, and can cause lung cancer if inhaled. The other type, which is Chrysotile, is less hazardous. Its fibers are silky in texture and curly. They do not get inhaled as readily as the other type, and the fibers are easier for the body to exhale.

Approximately 99% of asbestos used today is Chrysotile.

Also known as "White Asbestos," Chrysotile is made up of lots of curly fibers. It is part of the Serpentine group, and it is what gives Tiger Eye its chatoyancy (shimmer). Chrysotile in itself is usually chatoyant to a small degree.

The name "Chrysotile" comes from the Greek *chrysos*, meaning "gold," and *tilos*, meaning "fiber."

Chrysotile has been known for centuries, having been used in textiles such as oil lamp wicks and cremation clothes. However, it wasn't mined as asbestos until the 1800s.

Being that this stone is seldom used for jewelry or decoration, few medical and/or metaphysical properties have been attributed to it. However, it is said to help the bearer find his or her true self. Physically, it is thought to aid in problems with veins, arteries, and the pores of the skin.

As asbestos is known for its ability to resist fire and heat, the gem Chrysotile is thought to protect the wearer from fire, heatstroke, and sunburn.

Cinnabar

The bad news is that Cinnabar is highly toxic. It is a bright red ore of mercury; the main ore of this poisonous metal. Any beads made out of natural Cinnabar would be very dangerous to wear, as the body's heat would be enough to vaporize the mercury out of the stone and allow it to be absorbed through the skin.

The good news is that "Cinnabar" jewelry is not real Cinnabar. A polyester plastic resin is layered, molded, and then machine-cut or hand-carved into the desired shapes.

The use of Cinnabar (or fake Cinnabar) in jewelry and ornamental objects has its roots in the Mayan civilization as early as 2,000 B.C. The ancient Mayans would insert this deadly red stone into the death chambers of their loved ones, partly for decoration, but also to deter body thieves, who would avoid the Cinnabar.

Cinnabar was used to make the pigment known as Vermillion, used to color statues and faces in special ceremonies in ancient Rome. It was very expensive and used only in the most cherished of occasions.

The people of ancient China would make decorative carvings out of Cinnabar. These carvings were coated in lacquer to reduce their toxicity, but eventually the Cinnabar was left out of this process entirely, and replaced with red dyes.

The name "Cinnabar" may come from the Latin *cinnabaris*, which was their word for the mercury ore. It may also come from the Farsi (Persian) word *zinjifrah*, which means "dragon's blood" and which likely refers to the ore's deep red color.

Any jewelry vendor who claims to be selling real Cinnabar jewelry is either lying or trying to kill you. Be wary of such sellers.

Citrine

Citrine is a yellow or gold form of Quartz. The yellow color comes from a small amount of iron in the crystal, and it is actually rather rare. Most commercial Citrine is Amethyst that has been heat-treated, as Amethyst will turn golden when heated.

Known as the "Success Stone," Citrine attracts abundance, power, self-confidence, and self-discipline. It is known to be a happy, "cuddly" stone.

Citrine is also known as the "Merchant's Stone," due to the belief that it would bring wealth and prosperity. It is a good stone for business owners. Putting a Citrine in the cashbox is thought to attract money.

Citrine is used by crystal healers to stimulate the body's own healing energies, and is especially useful in digestive disorders.

Citrine was once thought to be useful in warding off poisonous animals and in treating bites from poisonous snakes.

The word "Citrine" comes from the French *citrin*, which means "lemon."

The darkest Citrine, a deep golden brown variety known as "Madeira Citrine," is the most valuable.

The ancient Romans used Citrine in jewelry and also in the art of intaglio, which is relief sculpting. Images were carved right into the surface of the stone for decorative art.

Citrine turns dark brown when exposed to X-rays, and white when exposed to excess heat.

Crystal healers appreciate that Citrine, carrying no negativity, never needs to be cleansed after metaphysical use.

Coral

Coral consists of the calcified skeletal remains of sea creatures called Coral Polyps. Their dense limestone skeletons cluster and build over many years to form coral reefs and islands.

Coral is very fragile and can be cracked and chipped easily. It is best cleaned in salt water. The most valuable Coral is the naturally red and orange types, though most red Coral on the market is dyed. Natural red Coral is highly valuable for its naturally brilliant color. The colors (natural or dyed) can fade in sunlight.

Black Coral is fairly rare in nature. The Coral is often black because it is in the early stages of decay. This is why black Coral colonies can take up to 50 years to mature.

As a substance organic in origin, it helps the wearer to connect with nature.

Coral has also been called "Bones of the Sea" and "Garden of the Sea."

Being from the sea, Coral was thought to carry the power of the oceans, and was carried by sailors to protect them at sea. It was thought to ward off hurricanes and bring good weather. It was once believed to cure madness, and was also worn to attract love and prosperity, and to ward off evil thoughts of ill-wishers. It was especially good for women (to promote fertility), children (to ease teething), and the elderly (to reduce the pain of arthritis). It was thought to protect children above all else, and is often given as a gift to children in many countries.

Modern healing properties of Coral include strengthening bone structure and healing mending bones, easing depression, to stop bleeding and to promote healing, and to protect against skin disease. It was once believed that red Coral would change color with illness. It would fade to white at the onset of illness, and would turn yellow with black spots as death approached. A dream of Coral was thought to foretell the recovery from a long illness.

crazy lace agate

Agates are a type of chalcedony Quartz that form in concentric layers in various colors. They form as round, lumpy nodules within rough, ugly crusts (geodes). Many beautiful agates are discovered in rivers after the water has eroded the outer crust away. The Crazy Lace variety consists of whorls and "eyes" of cream, tan, pink, brown and other colors. When dyed different colors, the "eyes" tend to stay natural, with the dye filling in the spaces around them.

Crazy Lace Agate is mined *only* in Mexico. Because of this, it is often called "Mexican Lace Agate" or "Mexican Agate." It is also sometimes simply called "Lace Agate."

Crazy Lace Agate is actually an Agate and a Jasper in one stone. Thus, it is called a Jagate or Jasp-Agate.

Crystal healers use this particular Agate to absorb emotional pain, to counteract physical low periods, to bring stamina to the wearer, and to gain insight to many options available in any decision-making process. It is thought to aid in sleep disorders such as insomnia and nightmares. It promotes laughter and humor, helping the wearer see life as a game. Because of its tendency to absorb pain from the wearer, crystal healers need to cleanse Crazy Lace Agate frequently. Cleansing can be done by placing the stone in sunlight, moonlight, or rain.

Physically, Crazy Lace Agate is thought to be good in healing problems of the skin, such as eczema and acne.

In ancient times, the "eye" patterns on Crazy Lace were believed to ward off the evil eye.

This stone is known as the "Generational Stone." It helps to bring grandparents and grandchildren to a closer understanding of each other. It is a great stone to wear when visiting with grandchildren, because it promotes calm in chaotic environments.

Dalmatian Jasper

This is a stone for those who want more fun in their lives. Dalmatian Jasper is said to help the wearer relax and enjoy life, and to be more open to fun situations and finding joy in everyday things. It is a good stone to carry with you when you expect to be under-stimulated.

While Dalmatian Jasper helps to find fun in everyday situations, it also helps the wearer to stay grounded in reality, and it lessens and removes disillusionment. This stone is helpful to relationships for the same reason; it enables the wearer to identify the strengths and weaknesses in relationships of all types.

Dalmatian Jasper is sometimes called "Dalmatian Stone," "Dalmatian Rock," "Dalmatine," and "Dalmatine Jasper."

The origin of the stone's name is obvious—it looks just like a Dalmatian dog. Like the dog, this stone promotes loyalty and is beneficial for long-term relationships. It also helps the wearer to regain trust that has been lost, and to diminish the need to take revenge on those who have done you wrong.

Dalmatian Jasper is primarily found in Mexico.

Healing properties include the ability to purify the blood, and protection from nightmares, depression, and negative thinking. It also helps to increase patience.

Dalmatian Jasper is also thought to protect against radioactive and environmental pollution. Old lore says that Dalmatian Jasper will sound a warning when danger is near.

Dalmatian Jasper is said to be a good stone for veterinarians and other people who work in the healing of animals. It calms animals and lets them know that the person is trying to help them. This stone also helps children to get over their fears of animals. It is thought to be a very protective stone for animals, so it is good to have around farms and zoos.

Diamond

Diamonds are the hardest mineral there is. Not only are they harder than anything else, but the degree of hardness is greater between Diamond and Corundum (the next hardest mineral) than it is between Corundum and Talc (the softest mineral). Some Diamonds are harder than others. For example, Australian Diamonds can cut South African Diamonds. Even though Diamonds are the hardest substance known on earth, they are also very brittle, so they can shatter under a hard blow.

One of the things that makes Diamonds unique is that they are always found as crystals. There simply are no non-crystal diamond specimens. This is likely because they are so hard that they cannot be eroded like other minerals can.

Another unique feature about Diamonds is that they are the only gemstone that are made up of one single element. Diamonds are pure crystallized carbon.

While Diamonds can come in several colors, the colorless variety is the most prized. Chemical impurities create the different colors, so a Diamond completely free of impurities will be perfectly colorless. The famous blue Hope Diamond, as well as other Diamonds with a smoky blue tinge, are not actually blue. The blue color is an effect caused by a complete lack of color paired with perfect transparency. These so-called Blue Nile Diamonds are the most prized of all.

If heated, a Diamond will decrease in size and finally disappear without any residue.

The word "Diamond" comes from the Greek *adamas*, which means "hardest steel" or "unconquerable."

The largest Diamond ever found was the 1 1/3 pound (3,000 carats) Cullinan Diamond, which was cut into nine large faceted gems and 96 smaller ones.

Diamonds are known from as far back as 400 B.C., when they were mined in India.

Diamonds are called "ice" because they are actually naturally cold stones, often several degrees below room temperature.

Diamonds are known as the "King Gems."

Approximately half of a rough Diamond's total weight is typically lost in the cutting process.

Hindu lore says that Diamonds were created when bolts of lightning struck rocks. They believe that Diamonds are very powerful, but will lose their powers if bought or sold. They must be given as a token of love or friendship in order to be used as a talisman.

Long ago, Jewish high priests would use Diamonds to determine the innocence or guilt of the accused. A Diamond was believed to grow dull and dark in the presence of guilt.

Ancient Romans said that Diamonds were splinters of stars. Ancient Greeks said that Diamonds were the tears of gods.

Even though we tend to set Diamonds in gold, they are actually thought to be most potent, powerful, and valuable if set in steel.

One of the most expensive gemstones in existence, Diamonds have been known to command over $925,000 a carat.

It is thought that stealing a Diamond will bring bad luck and catastrophic misfortune to the thief.

Dumortierite

Dumortierite is a very pretty stone, usually found in denim-blue but occasionally found in black and red as well. It is normally used for carvings and decorative items, and the lower-quality stones are used to color ceramics and porcelain.

Dumortierite was named for the paleontologist Eugene Dumortier.

Dumortierite is said to be a calming stone that helps the wearer collect his or her thoughts. It prevents the wearer from being scatterbrained; it restores order and organization in one's life. It promotes metal clarity to reduce excitability and stubbornness. It is a good stone for partnerships, as it facilitates tolerance and understanding of other people.

This stone is said to be helpful in combating addictions. It helps the wearer recognize repetitive patterns and to cope with eliminating them.

Dumortierite is often confused with Sodalite and other dark blue stones. Sodalite has more white, and is much lighter. Dumortierite is sometimes sold as imitation Lapis Lazuli.

While Dumortierite is usually found in its massive state (meaning that large, solid hunks are often found), it does have a much rarer crystalline form. The crystals are pleochroic, which means they change color depending on what angle from which you view them. The color changes from red to blue to violet.

Often found near sources of water, an African legend says that Dumortierite is petrified water.

Dumortierite is the main colorant in Blue Quartz (which is called "Dumortierite Quartz"). Tiny Dumortierite crystals grow in the Quartz, giving it its blue hue.

emerald

Emerald is the green form of Beryl. It is the same stone as the blue Aquamarine.

Called the "Stone of Successful Love," Emeralds are said to promote domestic bliss and to instill loyalty and sensitivity in couples. Thus, it makes a good gift for lovers.

The word "Emerald" originated from the Greek *smaragdos*, meaning "green stone," and much later was found in Middle English as *esmeralde*.

Due to the way the crystal grows, Emeralds tend to look best cut into a rectangle shape, known as the "Emerald Cut."

Crystal healers use Emerald to enhance memory and stimulate the use of mental capacity. It is said to be beneficial to the spine, lungs, heart, and the muscular system. It is also said to be soothing to the eyes.

Natural Emeralds almost always include some flaws. The way they grow is conducive to many small inclusions of other minerals (mainly of calcite), and this is why even good quality Emeralds look cloudy. Because the inclusions can make the stone fragile, almost all Emeralds are treated with oil to fill the microscopic cracks.

The earliest known Emeralds were mined in Egypt starting in 3,000 B.C., in a mine still known today as Cleopatra's Mines. This area only turns out poor-quality Emeralds today, though.

The ancient Egyptians called the Emerald the Stone of Spring because of its green color. They associated it with rebirth, and used it as an aid in fertility and childbirth.

The finest-quality Emeralds have been known to command higher prices than Diamonds of the same weight.

fiber optic cat's eye

This is a synthetic (man-made) stone. It is composed of a substance known as "Ulexite," which is spun glass fibers fused together and then machine-cut to form the shapes. Gem-quality Cat's Eye is actually a by-product of a material produced by the telecommunications industry. Tiny glass "hairs" are fused together into solid glass. Gem cutters cut the fiber-optic glass according to the way the fibers are aligned, in order to get the "eye" evenly spaced across the gem.

Any stone that displays the cat's eye effect, including Fiber Optic Cat's Eye, Tiger Eye, and Chrysocolla, to name a few, is said to have chatoyancy, which is a single streak of light that can be seen when the stone is rotated. Viewed from the side, Fiber Optic Glass is nearly transparent, no matter how thick the specimen.

Crystal healers put Cat's Eye into the same category as many other chatoyant stones. Even though the stone is man-made, it still contains the chemical properties of some types of natural glass (primarily Quartz), so it is said to posses the same metaphysical attributes. Cat's Eye brings serene happiness, it stimulates intuition and enhances awareness.

All Cat's Eye stones (natural or synthetic) are considered to be lucky for gamblers.

All Cat's Eye stones are said to protect from the Evil Eye and bring good fortune.

Fiber Optic Cat's Eye is the same material used in fiber optic telephone technology.

Being man-made and inexpensive, Fiber Optic Cat's Eye is a good substitute for naturally chatoyant gemstones, and makes good children's jewelry.

fire opal

While "fire" in Opals usually refers to the play of color across the surface, Fire Opals generally do not show any opalescence. The "fire" in this case refers to the color of the stone, which ranges from dark red to orange to yellow. This color is caused by iron impurities, and is unlike any other gemstone.

Fire Opals are the only Opals that are faceted instead of made into cabochons. They can be transparent or slightly milky, and occasionally show flame-like reflections when turned. Color play across the surface of the stone is very unusual, and any flash raises the value of the stone considerably.

Most Fire Opal is mined in Mexico, so the stone is often called "Mexican Fire Opal." Bright red Fire Opals are also sometimes called "Cherry Opals."

Fire Opals are generally lighter in weight than regular Opals.

Crystal healers use Fire Opals in the healing of blood disorders, and problems with the back and intestines.

This stone is thought to be useful in eliminating apathy, preventing burnout in stressful situations, freeing the spirit, promoting optimism and creative power, easing depression, and releasing deep-seated fears and grief. It helps the wearer deal with the past and let go of painful memories. When a person feels deeply mistreated or outraged, feeling that circumstances are not fair, Fire Opal will help the wearer to get through the shock and indignation.

Fire Opal was known to the Aztecs, and first introduced in Europe when it was brought back from the Americas by the Spanish conquistadors. The Aztecs and Mayans used it for jewelry and also for mosaics and in religious ceremonies. It was called *Quetzalitzlipyollitli*, which means "Stone of the Bird of Paradise."

Due to its origin and present location, Fire Opal is considered the National Stone of the Central American nations.

Fluorite

Fluorite comes in a very wide range of colors, with only Tourmaline featuring more colors. The most common colors are purple, green, blue, and clear, but Fluorite also comes in yellow, pink, red, and orange.

The name Fluorite comes from the Latin *fluere*, meaning "to flow." This refers to the fact that Fluorite was often used as a flux in smelting metallic ores, and also because it is very easy to melt.

Crystal healers use it for transferring certain types of negative energy into positive energy. Its mental healing powers make it a good stone to aid in sleep. It is believed to bring prophetic dreams.

Often called The Genius Stone, Fluorite is said to help amplify, focus, open, and expand the mind, creating new pathways for knowledge.

Fluorite's high percentage of fluorine is said to help strengthen teeth and bones, and to aid in the absorption of nutrients.

Fluorite exhibits two interesting physical properties. One is fluorescence (named for the stone), whereby the stone glows under a black light. The other is thermoluminescence, whereby the stone glows when heat is applied. This can be seen by taking a stone and heating it over an electric range in a dark room. However, this thermoluminescence is a one-shot deal; once it is seen, the glow will fade and will never again be seen in the same specimen. (One variety of Fluorite, called "Chlorophane," will glow simply from the heat of one's hand.)

Fluorite is an extremely soft stone (4 on the Mohs scale), so its use in jewelry is limited. It is more often made into carvings, but there are some beads and pendants made of Fluorite. Extreme care should be taken to avoid blows, as it cracks and scratches very easily.

Faerie Lore names Fluorite as a favorite stone of Faeries.

Freshwater Pearl

Natural Pearls form when a tiny irritant such as a bit of sand or grit finds its way into the shell of shellfish such as oysters or mussels. Over time (as many as ten years), the irritant is covered with layers of nacre, also known as Mother of Pearl, to form the roundish Pearl.

Long ago, Pearls were as valuable as real estate, because one would have to search thousands of oysters to find one single Pearl. These days, Pearls are "farmed" by placing a small bead into an oyster shell, and harvesting the completed Pearl a few years later.

Freshwater Pearls are formed in mussels rather than oysters. They are formed naturally in rivers and lakes, and cultured on farms.

There are three major things to look for in a Pearl:
1. Orient—the depth of the inner glow of the nacre layers
2. Shape—the rounder the Pearl, the more valuable
3. Color—Pearls can naturally occur in several colors, but most colored Pearls on the market are dyed

One way to tell if a Pearl is natural or man-made is to put it in your mouth and roll it against your teeth. A natural Pearl will be gritty, while a man-made Pearl will be smooth.

Pearls are known as the "Stone of Sincerity;" promoting faith, charity, innocence, integrity, truth, and loyalty. They are said to represent purity of body and mind.

Pearls are thought to inhibit rough behavior, and are a good stone for children to wear. They are also thought to keep children safe; as the stones of innocence and purity, they are the talisman of the innocent.

Pearls are very sensitive, and should be treated gently. Any perfumes or cosmetics should be put on an hour before donning the pearls, and after wearing they should be gently wiped with a soft cloth.

Fruit Quartz

The collection of "Fruit" Quartzes on the market today are not real Quartz. While each has its own name and color, the creation process is the same for all of them. Glass is heated to a liquid state, and dyes are poured into the liquid glass and slightly mixed, then allowed to cool and solidify before the colors can blend completely.

The most common of the Fruit Quartzes is Cherry Quartz, which has swirls of reddish-pink against the clear glass. The next two most common are Blueberry Quartz and Pineapple Quartz, which have blue and yellow swirls, respectively. (Pineapple Quartz is also often slightly iridescent.) Less-popular Fruit Quartzes are Lime Quartz (green), and Watermelon Quartz (pink). New fruit names are often given to variations on these main colors.

While most glass beads and components are molded—that is, the liquid glass is poured into a mold and allowed to cool in the desired shape—Fruit Quartz is made into large blocks, from which cabochons, beads, and faceted gems (as well as other shapes such as eggs and spheres) are cut. This gives the stone the look and feel of real gemstones.

Despite being man-made, each batch of Fruit Quartz has individuality just like real gemstones (as opposed to other glass gems such as Goldstone or Fiber Optic Cat's Eye, which have a uniform look throughout the stone). This makes it popular among gem cutters, who appreciate the individuality of each stone.

Two "Fruit" Quartzes that are actually real stones are Strawberry Quartz and Lemon Quartz. Strawberry Quartz is Quartz with needles of reddish stones such as Hematite, which give it a shimmery pink rutilated effect. Lemon Quartz is a type of pale Citrine.

Being man-made, Fruit Quartzes have no lore ascribed to them, though the pink color of Cherry Quartz is used for applications of romance, passion, and love, like many other pink and red stones. As glass, Fruit Quartzes are considered to have similar healing and metaphysical properties as natural glasses such as Obsidian.

garnet

Garnets are from a family of gems that come in every color except blue. The red-wine colored garnet is the most popular.

The word "Garnet" comes from the Latin *granatum*, which means "pomegranate seed."

Garnet has many metaphysical healing properties. It is used for cleansing blood, healing the heart, spine, and lungs. It is used to cure fever, and is used for balance and acceptance.

Long ago, people wore garnets as protection against insect bites, evil spirits, and the evil eye. Asiatic tribes once used garnets in place of bullets, believing their blood-red color was more deadly than lead bullets.

People used to believe that when danger was near, a garnet would lose its brilliance.

Plato had his portrait engraved on a Garnet by a Roman artist.

Garnets were once used as a symbol for lasting love, given when lovers would be apart, to ensure a quick return. This superstition had its roots in Greek mythology. Persephone, daughter of Demeter, the goddess of the harvest, once accidentally fell into the Underworld. Hades, god of the Underworld, wished Persephone to be his wife, but Demeter refused to allow anything to grow on earth until her daughter was returned to her. Hades eventually agreed to let Persephone go, but before this, he gave her a pomegranate. He knew that when she tasted the sweet fruit of the dead, she would return to him. She ate six seeds, and because of this, Persephone had to spend six months out of every year in the Underworld. During these months, Demeter was so distraught over her daughter being gone that she refused to let anything grow, and this was winter. The Greeks believed that Garnets represented pomegranates, and that a Garnet should be given to ensure one's lover would always return.

gold

Gold is the most highly valued metal in the world, in all cultures. Its beauty and usefulness is unsurpassed by any other. It has been known and used since prehistoric times, and its value has never wavered.

Gold is indestructible. Being completely non-reactive, it will not rust, corrode, or tarnish. It is 100% recyclable, and is regularly used in medical, industrial, and electrical applications. It is highly ductile (meaning it can be drawn into a very thin thread), and is so malleable that one ounce of Gold can be beaten into 300 square feet of Gold Leaf. It can be rolled out so thin that light will shine through it.

Gold is found naturally by itself, with no other elements contaminating it. It is usually mixed with other elements (such as copper or silver) to aid its usefulness, according to the intended purpose. Its purity is measured in carats. 24 carats is pure Gold; 18 carats is an alloy that contains 75% Gold.

Gold is the world's only naturally-occurring yellow metal. Its primary color is yellow, but is also sometimes found in black, red, and purple.

Gold is incredibly rare, and only comprises about five ten-millionths of the earth's crust.

The word "Gold" comes from the Old English word *geolo*, meaning "yellow."

Gold has been likened to perfection throughout history. The Egyptians used the perfect circle as a symbol for Gold. Early Greeks associated Gold with the Sun-god Apollo, known for his perfect beauty. Today's vernacular has many phrases using "Gold" as a euphemism for perfection: "Good as gold," "The gold standard," "Heart of gold," and many others.

While the 1849 Gold Rush makes us think of California as a big mining spot, most Gold in America is mined in Nevada and the Dakotas.

goldstone

Goldstone is *not* a real gemstone. It is glass with copper flecks in it.

Goldstone was created by accident in the European Renaissance period. Having been created *a venturi* ("by accident"), it was initially given the name "Aventurine Glass," but was later given the name of Goldstone.

Due to its sparkling copper flecks, Goldstone is sometimes called "Sand Stone," or "Red Sand Stone."

Other commercial names for Goldstone include "Sun Sitara," "Stellaria," and "Monk's Gold" and "Monkstone" after the monks who created it.

Goldstone was originally made by a group of monks in Italy, and to this day most of it is still made at this particular monastery (though it is made elsewhere too). The "recipe" was for a long time a very closely guarded secret, which was lost at some point, but has since been recreated. Due to its origins in the monastery, Goldstone is used frequently in religious (especially Catholic) jewelry.

In the making of Goldstone, copper salts are added to brown-colored liquid glass. As the glass cools, the salts turn into copper crystals.

The color of the stone is dependent on the color of the glass, not the crystals (which are always copper). Reddish Goldstone and dark Blue Goldstone are both popular and easy to find. Green Goldstone is *not* well heard-of and is very hard to find.

Because it is a man-made stone, it is hard to define any true crystal healing attributes of it. However, some people feel that it has the same powers as natural glass like Obsidian, and that the copper crystals in it provide great conduction of energy. Goldstone is said to promote calmness and stability. It is helpful in deflecting unwanted energies, which makes it a good general protection stone.

hematite

Hematite is an iron-rich (70% iron) stone that is a major ore of iron. Its metallic sheen hides its secret "true" color, which is blood red. When Hematite is sliced very thin, the slices are red and transparent. In powdered form, it is blood red and is used for pigments (and is the ingredient of the Red Ochre that some Native American tribes use for face paint).

Hematite is usually found in globular (round, bumpy) form; actual crystals are very rare and sought after by collectors.

Crystal healers use Hematite for any disorder of the blood, as well as for leg cramps and insomnia. It is thought of as the "Stone of the Mind," bringing clarity of thought to the wearer, helping to focus the mind, enhancing mental capabilities, enhancing memory, and promoting original thinking. It is said to be a calming stone, and helps to allow the wearer to "Reach for the Stars" and understand that personal limitations exist only in the mind.

The word "Hematite" comes from the Greek *haima*, meaning "blood."

Hematite is thought to protect against ionized radiation, and is therefore a good stone to wear when working around computers.

There are imitation Hematites on the market, called "Hematine and "Hemalyke."

Hematite often forms around areas of standing water or hot springs. When large quantities of Hematite were found on Mars (which looks red due to its high iron content), this caused scientists to theorize that Mars had once had water on it.

Long ago, polished Hematite was used as mirrors.

Legend has it that Hematite formed in the earth during battles, when the blood shed during fighting fell to the earth.

hemimorphite

Hemimorphite is an ore of zinc. Occurring alongside the similar mineral Smithsonite, the two minerals were initially thought to be the same, and they were collectively called "Calamine." Later, it was found that Calamine was actually two minerals.

Hemimorphite and Smithsonite are often ground up and made into the zinc oxide solution called Calamine Lotion, which is used to sooth skin irritations. In addition to skin applications, crystal healers use this stone to reduce stress, ease feelings of anger, and help all areas of brain function, including brain injuries, dyslexia, and coma.

Hemimorphite is found all over the world, but some of the best deposits are around the border of Belgium and Germany. It was known in Ancient Rome as "Galmei" or "Cadmia."

A generally brittle, soft stone with lots of hairline fractures, Hemimorphite is rarely used in faceted jewelry. It is much better used as cabochons and beads, which are stronger and can withstand daily use as jewelry. Transparent crystals are never found larger than three carats, but in its massive form specimens of over 1000 carats have been found.

Hemimorphite can come in white, gray, yellow, and brown, but its most common color is blue or green-blue. It gets this color from impurities of copper. It is sometimes dyed to enhance the blue color.

The stone gets it name from its strong hemimorphism (Greek *hemi*, meaning "half," and *morph*, meaning "shape."), wherein its crystals have different shapes at either end. One end of the crystal is rather blunt, while the other is shaped like a pyramid. This is a very rare occurrence in crystals, and while some other stones show it, none show it as strongly as Hemimorphite.

The stone's crystal shape causes it to be pyroelectric (susceptible to electric charges when heated), and piezoelectric (creating electricity with changes in pressure). This often causes dust to build on specimens.

Herkimer Diamond

Herkimer Diamonds are naturally occurring double-terminated Quartz crystals. While most Quartz grows on a host rock, needing to be broken off, Herkimer crystals grow in a very soft matrix of Dolomite, and usually have little to no contact with the host rock at all. Thus, they are terminated at both ends, and can be easily removed from the host rock without breaking.

Herkimer Diamonds are not real diamonds. They are given this name because of their perfect 18-faceted shape and unusual clarity. Most Herkimers are water-clear, though some have inclusions. (There have also been some Amethyst Herkimer Diamonds found.)

These crystals are found in Herkimer County, New York. Similar crystals are found elsewhere, but they are not official "Herkimers."

Herkimer Diamonds are slightly harder than natural Quartz—about 8 on the Mohs scale. They carry the same metaphysical healing properties as Quartz, but it is magnified with Herkimers, due to their exceptional clarity. The metaphysical properties of Herkimers are thought to be the strongest in existence.

Crystal healers use Herkimer Diamonds as a filter to release toxins from the body, to aid in clarity and creativity, to boost the immune system, and to enhance the dream state.

Known as the "Stone of Attunement," Herkimers attune the user to other people, environments, and activities.

Mohawk Indians knew of Herkimers, as did the early European settlers. However, the first Herkimer Diamond mine would not be opened until the 1950s, when a farmer, having found large numbers of crystals on his property, opened the mine to prospectors for one dollar a day.

hessonite

Also known as "Cinnamon Garnet" and "Champagne Garnet," this is a variety of Grossular Garnet that is brown, orange, red, or peach colored. (The green variety of Grossular is Tsavorite.)

From a distance, Hessonite sometimes looks bright red, and its color can brighten when viewed in artificial light as opposed to natural sunlight.

Almost all Hessonite comes from Sri Lanka, though there are some much smaller deposits in other areas of the world.

While some Hessonite gems are transparent, most contain a type of inclusion that give the stone a cloudy, swirled look. These inclusions are called "treacle," for their resemblance to the syrup-based British candy of the same name.

Hessonite is thought to aid in letting go of outgrown ideas, views, habits, and behaviors of the past. It promotes positive change in the wearer's life. It encourages new challenges, and provides the courage to accept those challenges.

Called the "Creativity Stone," Hessonite stimulates creativity in all forms. It aids in self-respect and respect of others, and eliminates feelings of inferiority that may lead one to give up creative tasks. It is considered a strong spiritual stone for wise people; it prevents gossip-spreading and gives the ability to approach troubles with intelligence.

Physically, Hessonite is used to regulate hormones, stimulate metabolism, bolster the immune system, and promote good health through better absorption of nutrients in foods.

The word "Hessonite" comes from the Greek *hesson*, meaning "slight," which refers to the fact that it has a low specific gravity.

howlite

Howlite is a dullish white or light gray stone with gray marbling throughout. Also called "White Buffalo Stone," it occurs in large nodules that look like cauliflowers.

A very soft stone, Howlite takes dyes easily, and is therefore often used as an inexpensive substitute for the light blue Turquoise (called "Turquonite" or "Turquanite"), dark blue Lapis Lazuli, purple Sugilite, or Red Coral. When it is polished, it is beautiful as a gem-quality stone on its own merits, without dye.

Howlite is named for Henry How, the mineralogist who first discovered it in Nova Scotia.

Found all over North America, it is mined mainly in California, where nodules of up to 100 pounds have been found.

Howlite is thought to relieve stress, and to help alleviate insomnia.

Howlite is an exceptional stone to help promote goodness and banish bad behavior. It encourages decency, politeness, patience, tact, and character, while discouraging cruel behavior, selfishness, critical attitudes, and rudeness. It absorbs anger—either your own anger, or someone else's anger directed toward you.

Howlite also promotes creativity and imagination, improves memory, produces a thirst for knowledge, and aids in achieving artistic goals. It is a good stone for an artist to wear.

Physically, it is thought to help strengthen teeth and bones.

Howlite is sometimes worn by children to help them be less afraid of the dark.

hydrogrossular garnet

There are several families of Garnets. One of these families is the Grossular group (also called Grossularite), which includes Hessonite, Tsavorite, and several others. The Grossular group comes in colorless (which is very rare), red, orange, yellow, brown, green, and black. It is the most colorful group of all the Garnets. The Garnet discussed here is a Grossular Garnet known as Hydrogrossular. It is a generally opaque stone in shades of yellow, green, and brown.

Hydrogrossular Garnet is sometimes called the "Gooseberry Stone" due to its resemblance to gooseberries. It is also colloquially known as "African Jade," and makes a good substitute for Jade in color and workability. (It is seldom faceted; it works better as cabochons and carvings like Jade.) It is also sometimes called "Transvaal Jade," which is its official trade name.

One of the unique characteristics of Grossular Garnets is that they contain tiny crystals (usually of the mineral Diopside), which, when viewed under a microscope, give the stones a particular swirled pattern known as "treacle." This is one way to positively identify a stone from the Grossularite group.

The word "Grossular" comes from the Greek *grossularia*, meaning "gooseberry."

Crystal healers use Hydrogrossular Garnet to treat problems of the kidneys and intestines.

Hydrogrossular Garnets are thought to help solidify partnerships; thus, they are good to use in business partnerships. It is also a good stone for husband and wife to give to each other. It helps to keep long-distance relationships (whether friendship, romance, or business) going strong. It is thought to help keep the wearer cool in times of crisis.

Iolite

Iolite is a generally blue-violet stone, but shows extreme pleochroism, which means it looks different colors when viewed from different angles. In a cut Iolite gemstone, the stone looks dark when viewed from above (down the crystal axis), and lighter when viewed from the side (across the crystal axis). In a cube-shaped Iolite, it may look as dark as a sapphire from one side, clear as water from the other side, and honey-yellow from the top.

Iolite's extreme pleochroism gave rise to its alternative name, "Dichroite."

Iolite is one of the most difficult stones to cut; not because of its hardness, but because of its pleochroism. The stone must be cut in a certain direction to take the best advantage of the color-change properties, and this can be hard when the shape of the rough stone doesn't flow well with the pleochroism.

Iolite has a very important place in history as the world's first polarizing filter. Leif Eriksson and other Viking mariners would use it in their journeys as a rough compass. The stone's pleochroism enabled them to determine the sun's location on overcast days. Looking through an Iolite lens, they could tell the position of the sun (the lens would look bluest at 90 degrees from the sun).

Iolite is also known as "Water Sapphire," a reference both to its color and to its ancient usefulness at sea. Today, it is thought to protect the wearer on marine journeys.

The name "Iolite" comes from the Greek *ios*, meaning "violet," and *lithos*, meaning "stone." The scientific name is "Cordierite," after the French geologist P. Cordiere.

Crystal healers use Iolite to help in dealing with addictions. It assists in detoxifying the body and squelching impulses.

Iolite is also thought to enhance curiosity, and to help in building relationships.

Jade

"Jade" is an all-purpose name that encompasses two main types of gemstone: Jadeite and Nephrite. Jadeite is the rarer of the two, and comes in a variety of colors, including white, green, red, blue-green, brown, purple, and yellow. The highly prized deep green "Imperial Jade" is a form of Jadeite. Nephrite is more common, tends to stay in the green range, and is often used for carvings due to its smoothness. Jade is often dyed to enhance its color.

A sacred stone to the Chinese and Mayans, Jade is said to bless everything it touches. Often found in religious carvings, it is thought to be the symbolic link between mankind and the spiritual world, and is highly prized for spiritual growth. It promotes wisdom, peace, harmony, and devotion to one's higher purpose. It also promotes confidence, self reliance, and self-sufficiency, and aids in building dreams into realities.

The Chinese call Jade the *Yü*, which is a word that applies to all precious things. It is the Chinese Royal Gemstone.

Also known as the "Dream Stone," Jade placed under the pillow at night will promote lucid dreaming. It is also said to be the concentrated essence of love, and is a good stone to give to a lover. Jade's ability to promote calm in chaotic environments makes it a good stone for parents of active children to carry.

Ancient Chinese lore says that Jade was created during a storm, and this is why Chinese households always have Jade in them to protect them from lightening. Another legend says that Jade was crystallized moonlight, brought to earth from the mountains.

The mythical phoenix of Chinese and Japanese lore was said to land only on surfaces of Jade.

The stone known as "New Jade" is actually a non-Jade stone called Serpentine. It looks a lot like light, sea green Jade, and hence it is called Jade.

Jet

Jet is a black stone that is comprised of a coal called Lignite. It is from ancient trees that burned and fossilized. It is the only gemstone that is naturally pure black. It is sometimes called "Black Amber," because it is organic like Amber and comes from trees, but it is not related to Amber.

Due to its composition of hydrogen, oxygen, and carbon, Jet can be burned. In ancient times, it was burned and the smoke used to ward off evil spirits.

The word "Jet" is from the French *jaiet*, which is the French word for this material.

Jet is a fairly soft material, but lends itself well to carving. In Victorian times, Jet was often made into intricately carved beads and buttons. Back then, it was believed to be a good stone to wear in mourning, as it was thought to bring grief to the surface to be healed. In the 1900s, jewelry made from Jet was deemed suitable for mourning, due to its dark color. It was also used to protect against the evil eye, and against thunderstorms. To dream of Jet was thought to foreshadow upcoming sorrow.

Jet can sometimes have Pyrite inclusions that give it a shimmery golden glow.

Metaphysical healers recognize the power of Jet as a natural organic material. It is thought to enable the wearer to tap into the ancient wisdom and power of Mother Earth, and to draw power and knowledge to the wearer. It is also thought to stabilize finances.

Jet was also often used to make rosaries for monks. In the 1920s, long strands of Jet beads were worn by women.

Like Amber, Jet has an electrical charge, and will pick up little bits of paper when rubbed.

Kambaba Jasper

Kambaba Jasper isn't truly a Jasper; it is a sedimentary rock found on Madagascar, off the coast of Africa. Its main colors are dark and mint green swirled together. It is relatively new to the gem market.

This stone is highly prized for its exotic colors and patterns, which are created by the fossilization of algae in sedimentary limestone deposits. It is one of the oldest fossils known. The patterns bring to mind a rainforest as seen from above. It is an excellent stone for making into cabochons.

Kambaba Jasper is also known as "Kambaba Stone," "Stromatolite" (due to the stone being made of fossilized stromatolite algae), and "Crocodile Jasper" (due to the stone's resemblance to amphibian eyes).

Crystal healers use Kambaba Jasper to aid in dietary stabilization. It helps the body to assimilate vitamins and minerals from foods, and it helps to cleanse the body of toxins.

It is also thought to aid in the healing process after an illness or injury.

Kambaba Jasper is said to soothe the nerves and calm the mind. Known as the "Supreme Nurturer," it promotes a feeling of peace in the wearer. It brings wisdom and the ability to get along with others, diminishes worry, and provides support through times of stress.

Kambaba Jasper is thought to be beneficial for plant growth and health, particularly in arid areas with poor soil.

There has been some controversy with unscrupulous sellers trying to pass off Kambaba Jasper as the more famous and sought-after Nebula Stone. But Nebula Stone is an igneous rock (composed within the earth), while Kambaba Jasper is a sedimentary rock (composed of cemented sediments).

This stone is thought to be a useful aid in connecting with ancestors.

Kunzite

Kunzite is a relatively new stone, discovered in the early 1900s. It is the lilac and pink version of Spodumene (the green version of Spodumene is called Hiddenite), and is very rare and expensive (it is not uncommon to see prices of $10.00 or more for a single bead).

While Kunzite is fairly hard—about the same hardness as Quartz—it is extremely fragile. Its perfect cleavage (planes of weakness) make it hard to cut. If it is hit hard, it can fracture along the cleavage lines. Kunzite should never be left near overly hot areas, it should be protected from scratches and blows, and will fade if exposed to sunlight. (For this reason it is dubbed the "Evening Stone.")

Kunzite is pleochroic, which means it appears to be different colors when viewed from different angles. From one angle it looks clear, while from another it looks pink. Gem cutters have a hard task of cutting it so that it looks pink from above. Smaller stones don't look very pink when cut; a gem must be 10 carats or larger for a good pink color.

Physically, Kunzite is thought to aid in problems of the heart, nerves, circulatory system, and lungs.

Emotionally, Kunzite is thought to bring peace and inner calm, to help the wearer to focus on the task at hand, and to remove mental obstacles and dissolve negative emotions. This stone helps to foster self-discipline, and to break habits and addictions. It promotes love in all forms, and brings the wearer a sense of self-respect and inner strength. Its high lithium content makes it an excellent stone for aiding in depression.

Kunzite was named after American geologist George F. Kunz, who first described it.

Kunzite (and its green cousin Hiddenite) crystals can grow to enormous sizes. The largest crystal on record, found in South Dakota, was 75 metric tons and 42 feet long. The largest faceted Kunzite is 880 carats.

Kyanite

Kyanite is a very unusual stone in that it has two hardnesses on the Mohs Scale, depending on the direction it is tested. It grows in crystals shaped like long, flat blades, and it cleaves along perfect lines. Whereas other minerals have the same hardness all around, Kyanite shows a hardness of 4-5 when tested lengthwise, and 6-7 when tested crosswise.

While Kyanite is typically a blue color, it sometimes comes in green, white, and yellow. It exhibits strong pleochroism, which means it looks different colors when viewed from different angles. Its color can look dark blue from one angle, and white or colorless from another.

The word "Kyanite" comes from the Greek *kyanos*, which means "blue," in reference to its color. This mineral is also known as "Disthene," which comes from the Greek *di*, meaning "two," and *stenos*, meaning "hardness," a name given because it has two hardnesses.

Metaphysical healers use Kyanite as a calming stone. It helps to focus the mind and dissolve mental and emotional confusion. It helps the wearer let go of his or her anxiety. It keeps the mind focused and able to ward off distractions.

Kyanite is thought to be a highly artistic stone, enabling the wearer to tap into his or her creative talent in art, music, dance, writing, and other artistic expressions.

Gem-quality Kyanite is extremely rare. Its perfect lines of cleavage and its brittle, splintery nature make it very difficult to cut, and stones of fine color and transparency are rare. However, its blue color and clarity when faceted rivals more precious gems such as Sapphire and Aquamarine, but at a fraction of the cost.

Commercially, Kyanite is used in the manufacture of spark plugs and heat resistant ceramics.

Labradorite

Labradorite is a type of Feldspar that occurs mainly in white, yellow, and gray, and dull green. Its main characteristic is its shimmering "labradorescence," which is a blue, yellow, or green sheen under the surface of the stone. This is caused by thin, papery layers (called lamellae) within the stone. Light enters the stone, is reflected against all the many layers, and leaves the stone at a much slower wavelength, causing a beautiful sheen across the outer surface of the stone.

Labradorite is sometimes called "Black Moonstone," and is very closely related to Rainbow Moonstone.

Labradorite gets its name from the location of its discovery, along the coast of the Labrador Peninsula in Canada.

The finest quality Labradorite is found in Finland, and is commercially called "Spectrolite."

Crystal healers use Labradorite to clarify the eyes, cure disorders of the brain, and promote good digestion and metabolism. Its metaphysical uses include sharpening mental acuity, reducing stress, and promoting calm amidst chaos.

Labradorite is said to bring out the best in the wearer, and is sometimes known as the "Self-Esteem Stone."

Called the "Wizard's Stone" in mystic circles, this stone was worn as a good luck charm in the 18th century, and is still believed to be a source of good luck. When worn around the neck in particular, Labradorite is thought to bring good luck in love and health.

In Eskimo lore, the Northern Lights used to be trapped in the rocks along the coast of Labrador. One day, a wandering Eskimo found them and released them into the air with a blow from his spear. However, some of the lights remained trapped in the rocks. These rocks are what we know as Labradorite.

Lapis Lazuli

Lapis Lazuli (also know simply as Lapis) is a rock comprised of many minerals. The main mineral is Lazurite, and it can also contain Calcite (which gives it white streaks), Sodalite (which gives it its blue color), and Pyrite (which gives it golden flecks), as well as a number of other minerals. As a rock, rather than a typical gemstone, Lapis can have slightly different mineral compositions from stone to stone.

In general, the fewer white lines, the more valuable the stone. The most expensive Lapis is a deep, solid blue with small Pyrite inclusions. Sometimes the white areas of Lapis are colored blue to raise the value of the gem. Lapis chips and scratches easily and should be protected.

True Lapis is hard to come by. Most "Lapis" on the market is actually dyed Howlite or some other stone. Pyrite inclusions in the stone are usually indicative of real Lapis.

Lapis is one of the oldest gemstones known. Ancient Egyptians considered it a sacred and prized stone. Looking like the star-studded night sky, it was thought to possess the energy of the sky, limitless in its wisdom.

Ancient Egyptians used powdered Lapis as eye shadow. Powdered Lapis was also mixed with oil to form the pigment known as Ultramarine, which was used in paintings for thousands of years.

Both Christians and Egyptians associated Lapis with motherhood. Hindus believed that a Lapis stone on a gold chain would protect children from evil.

The stone's name comes from the Latin *lapis*, which means "stone," and the Farsi (Persian) *azul*, which means "blue."

Crystal healers use Lapis to reduce fever and sore throat. It is thought to help the wearer understand his or her emotional issues; not necessarily reducing them, but bringing them into a clear light so that the wearer has the opportunity to understand and deal with them.

Larimar

Larimar is the gem name for the rare blue variety of the mineral Pectolite.

Larimar is volcanic in origin. Erupting magma mixed with assorted minerals to form the blue Pectolite.

Found in only one place in the world—the island of Hispanola in the Dominican Republic—Larimar is an incredibly rare stone, and as a result is very expensive.

This is a relatively new stone. Originally discovered in 1916 by Father Miguel Domingo Fuertes Loren, the proposed mine was never explored because the Ministry of Mining didn't know what the stone was. Later, in 1974, a visiting American named Norman Riling, along with a Dominican local named Miguel Méndez, rediscovered the mine, and convinced the Ministry to mine there.

Larimar got its name from a combination of Larissa (Méndez's daughter's name), and *mar*, the Spanish word for "sea."

The quality and value of Larimar depends on its color. White and light blue stones are the lowest grade, whereas brilliant blue stones are most valuable and treasured. The most highly valued stones are the rarest ones which show chatoyancy (a cat's eye effect).

Larimar is photosensitive, which means that its blue color will fade over time if exposed to bright light.

Crystal healers use Larimar as a calming stone. Its smooth blue color inspires tranquility and peace. It aids in mental clarity and love.

Physically, Larimar is thought to be helpful in disorders of the thymus, thyroid, and immune system.

Larvikite

Larvikite is a grayish type of rock called Feldspar, which has some bluish highlights. The blue can occasionally look pearly or iridescent due to crystals in the stone. It is related to Moonstone and Labradorite, and shows a lot of the same "fire" flashes that these stones do.

This stone is used primarily for exterior and interior facings in buildings. Great slabs of shimmering Larvikite adorn counters of pubs and walls of churches all over Europe. Despite its beauty, it is rarely used as jewelry, though recent years have found it being used more and more in beads, cabochons and carvings.

Larvikite comes from Scandinavia. It is also known as "Norwegian Pearl Granite," "Blue Pearl," "Norwegian Moonstone," "Pearlspar," and "Black Moonstone."

While this stone is new to jewelry making, crystal healers have already found uses for it. Larvikite is thought to help enhance youth and vitality. It aids the brain in taking in new information easily; it helps promote change.

Physically, Larvikite is thought to calm nerves and reduce blood pressure. It is an all-around calming stone.

The name Larvikite comes from the Larvik Fjord Region of southern Norway, where it is mined. Geologists believe that Larvik is the only natural location for this stone. Smaller deposits have been found elsewhere in the world, but these are thought to be the work of human intervention.

Due to the presence of magnetite in Larvikite, it can sometimes be slightly magnetic.

One of Larvikite's alternate names is "pub stone," because it is a common facing in British pubs.

Lepidolite

Lepidolite also goes by the name of "Lavenderine," due to its light purple color. It is more rarely yellow as well.

Lepidolite is an uncommon form of the common mineral, Mica. Mica is very soft and tends to flake easily into very thin panes.

Lepidolite contains a large percentage of lithium, and is a main source for lithium.

The pretty purple-pink color comes from the presence of Rubellite, which is Pink Tourmaline.

The name "Lepidolite" comes from the Greek *lepidos*, which means "scale," and *lithos*, which means "stone." The stone's thin flakes look scaly. Before it was officially named, it was known as "Lilalite," from the Hindu word *lila*, meaning "play."

The presence of lithium causes Lepidolite to be used as a healing stone for stress, depression, and bad dreams (especially those caused by stress). It is sometimes known as "The Dream Stone," believed to eliminate nightmares. It is also used to heal muscle pain.

Lepidolite has only recently been considered for a gemstone. The flaky nature of the stone causes it to be hard to cut, so it tends to be expensive. Its beautiful sparkly lavender color makes it very attractive in jewelry, though. All Lepidolite jewelry must be treated with extreme care, because it is soft enough to scratch with a fingernail.

When heated, Lepidolite colors the flame red, and fuses to a white glass.

Lepidolite is commonly found in Brazil, Russia, and parts of Africa. It is also found in large amounts in California.

Lodalite/Lodolite

Lodalite was named from a Greek word *lodos*, which means "mud" and *lithos*, which means "stone." The general translation is "Stone from mud." This is because Lodalite is thought of as "dirty Quartz." It is clear Quartz with inclusions of assorted minerals, including Chlorite, Iron Oxide, Calcite, and Hematite. The inclusions come in many colors, and give the stone the look of garden landscapes or underwater scenes. Thus, the stone is sometimes called "Landscape Quartz," "Scenic Quartz," and "Garden Quartz." It is also simply called "Inclusion Quartz."

Lodalite is a fairly rare stone and is mined in only one location in the world, Minas Gerais, Brazil.

Cut into cabochons with the bulk of the inclusions at the bottom of the cab, and a clear dome of Quartz above it, gives the impression that the viewer is looking down through a crystal ball into a different world. Because of this unique beauty, Lodalite has become very popular in recent years, and the supply is in danger of depletion.

Crystal healers love this stone for its ability to enhance extrasensory perception (ESP), and to heighten knowledge of past lives. It helps the user to get into a meditative state.

In addition to having all the metaphysical and healing properties of Quartz, Lodalite also contains the properties of its various inclusions. So when using it as a healing stone, crystal healers must determine which minerals are found within the Quartz.

Known as the "Stone of Power," Lodalite helps to amplify energy. It is a good stone to use in conjunction with other stones; it increases the power of the other stones that are used with it.

Lodalite is sometimes called the "Seer's Stone" or "Shaman's Stone." It is used as a shamanic gazing crystal. Looking through the clear Quartz to the depths of the inclusions is said to enhance meditation.

Mahogany Obsidian

When volcanic lava cools, it doesn't have time to form crystals, so it cools into a sort of natural glass. This glass is often black, but a high concentration of iron in the lava can make it a reddish-brown color. When two lava flows of different colors cool side by side, the result is a streaked rock. Mahogany Obsidian is the result of iron-rich lava (red-brown) that cooled with regular lava (black).

Small shards of Mahogany Obsidian (formed when droplets of lava cooled in the air after an eruption) found in the San Carlos Apache Reservation in Arizona are called "Apache Tears."

Almost all of the current sources for Mahogany Obsidian in the United States today were discovered by Native American tribes. Early Americans used this sharp stone as arrows and spears. Ancient Incas used it for mirrors, weapons, and masks.

Ancient Greeks associated Mahogany Obsidian with Hades, the god of the Underworld.

Crystal healers use smooth, cool, glassy Mahogany Obsidian to increase sensuality, particularly the sense of touch. It stimulates the growth of physical, spiritual, intellectual, and emotional centers, and provides strength in times of need.

Physically, Mahogany Obsidian is thought to be good for the gums.

This stone promotes environmental awareness and a desire to care for the earth. It helps the wearer to connect to planet Earth, and to help encourage Earth healing.

malachite

Malachite is a copper carbonate, with copper giving it its distinctive deep green color and exotic banding. It was once melted to make copper.

Malachite forms in lumpy masses rather than crystals. It is used primarily for carvings and cabochon jewelry. Most Malachite jewelry is fairly expensive.

The word "Malachite" might come from the Greek *malakos*, meaning "soft" (to refer to Malachite's hardness), or from the Greek *malache*, meaning "mallow," which is a plant of this bright green color.

Crystal healers love Malachite for its many healing properties. Among other things, it is believed to aid digestion, heal injured muscles, ease childbirth, ease colic, protect against radiation, and help with heart problems. Emotional benefits include the ability to assist the wearer in changing situations, promotion of fidelity in love and friendship, promotion of loyalty in partnerships, and the ability to clear emotional blocks.

Malachite amplifies every aspect of feeling, including the negative. Thus, it is an unwise choice to wear when you are in a bad mood.

The ancient Egyptians used Malachite in jewelry, and also ground it up to make cosmetics. In the Middle Ages, parents attached Malachite to their children's beds to ward off evil spirits and witches.

Malachite is said to be the "Guardian Stone of Travelers," and is thought to protect pilots in particular.

Due to its hardness of only 3½ on the Mohs scale, Malachite is a rather fragile stone that should be treated with care. It tends to scratch easily, and shouldn't be subjected to strong blows or extreme temperature changes.

Old lore says that Malachite will shatter into pieces if danger is impending.

mookaite

Mookaite is almost exclusively mined in Australia. It is found primarily on Mooka Station, a sheep ranch in Western Australia. The Aboriginal word *mooka* means "running water," and probably refers to the headwaters that feed the nearby Mooka Creek (though some theorize that the creek itself was actually named for the rock that is found under it).

Mookaite, also spelled "Moukaite" and "Mookite," is the common name for the stone Windalia Radiolarite. It is also sometimes simply called "Mook Jasper."

Mookaite is unofficially called a Jasper, but it is actually a fossiliferous sedimentary rock. It has mainly red and yellow colors, and a few other secondary colors. The stone often shows very interesting swirled designs, and sometimes shows evidence of once-living creatures. Occasionally the stones have dendrites (branch-like crystals), which look like little black trees in the stone.

In Australia, Mookaite was and still is considered a powerful healing stone. Its main purpose is to promote taking chances and accepting chance happenings, and to encourage the desire for variety and new experiences. It promotes inner calm while simultaneously promoting the desire for adventure.

Physically, it is believed to promote good health by stimulating the immune system.

Mookaite is thought to slow or stop the process of aging. It helps the wearer readjust his or her attitudes about aging, thus slowing the aging process through changing perceptions. It helps the wearer remain "young at heart" as well as physically young.

Mookaite is often made into jewelry for children, under the belief that it helps to modify children's behavior. It also helps children settle into a new home or other new environment.

moonstone

Moonstone, so named because of its resemblance to the shimmering moon, comes in several colors (white, yellow, blue, pink), with the best specimens being of a milky-white color. Rainbow Moonstone exhibits a bluish adularescence (luminous reflections or "fire"), and other Moonstone exhibits chatoyancy (cat's-eye shimmer).

Moonstone is a feminine stone, and has traditionally been associated with goddesses, especially moon goddesses. It's called the "Mother Earth Stone," and is helpful for all manner of feminine situations, including emotional balance, menstrual difficulties, hormone imbalances, menopause, and childbirth.

Moonstone is said to soothe emotions, guard against worry, and help the wearer to trust his or her intuition.

Known as the "Traveler's Stone," it is thought to protect the traveler from the perils of journey. It is considered a safe stone for sea travel, as well as for night journeys.

In India, Moonstone is considered sacred, and is known as a talisman of good fortune. Hindu legend says that Moonstone was formed from moonbeams.

Lore says that two people wearing Moonstone when the moon is full will fall immediately in love. It is a good stone for lovers, thought to arouse tender passion between a man and a woman.

It was once thought that carrying Moonstone in the mouth during a full moon would allow the user to see into the future.

In 1969, when humans first walked on the moon, the state of Florida (from where the moon launch had originated) wanted to claim Moonstone as its official gemstone. A year later, this request was granted—though there is no Moonstone to be found in Florida, nor is it found on the moon.

Moqui Marbles

Moqui Marbles are concretions of iron. They consist of a spherical sandstone center surrounded by Hematite. Sometimes hematite rings can be found within the ball. They are found when the Sandstone surrounding them wears away, leaving the harder iron behind.

Moqui Marbles are also called "Shaman Stones," "Mochi Marbles," "Ironstone Concretions," and "Navajo Cherries."

Formed 190 million years ago in the Navajo Sandstone Formation in southern Utah and northern Arizona, Moqui Marbles are found today where the formations have weathered and formed sand dunes. As the concretions weather out from the Sandstone, they are tumbled and polished by shifting sand. The outside of the balls can be very rough, or smooth.

Moqui Marbles can be a few millimeters in size, or up to five inches in diameter.

Moqui is a Hopi word that means "dearly departed ones." Ancient Hopi legend says that the spirits of ancestors came out at night to play with the marbles. Today, putting a Moqui Marble in the home is said to welcome the spirits of departed loved ones.

Tribal shamans use Moqui Marbles in their rituals by throwing them into the fire, where they explode in a shower of sparks.

Moqui Marbles are thought to be the most energetic stones on earth. They are used for cleansing, relaxation, and pain relief.

Wearing a Moqui Marble is thought to keep away false friends.

Today, Moqui marbles are protected in Utah's Grand Staircase-Escalante National Monument. In other areas it is legal to collect them.

moss agate/ tree agate

Moss Agate and Tree Agate are clear to milky-white forms of Quartz with impurities of hornblende growing in fissures in the stone. Technically, it is not a real Agate. It has the same chemical composition as Agate, but the design occurs in pebbles and fissures rather than in layers.

When the surrounding rock is mainly clear, the stone is called Moss Agate. When the rock is mainly white, the stone is called Tree Agate.

While the impurities do resemble moss, this is actually *not* where the name came from. It was first made popular near the seaport of Mocha (Al Mukna) in Yemen, and so it was called "Mocha Stone" and "Mocha Agate," which later got changed to Moss Agate.

Crystal healers use Moss Agate as a cooling stone—to reduce fevers and inflammations, to quench thirst, and to ease stomach acids. It helps to alleviate the symptoms of colds and infections.

In some cultures, powdered Moss Agate mixed with water was thought to cure insanity.

Moss Agate is thought to bring peace and patience to those who have extra chaos and stress in their lives. It balances male and female energies when they become extreme, and it helps the wearer achieve personal growth. It also helps to balance the physical and emotional energies in the wearer.

Moss Agate is thought to be a good stone for pregnancy, childbirth, and problems of the female reproductive system.

Moss Agate is thought to help in the growth of plants, and is known as the "Gardener's Stone." It is a good stone for use by botanists and farmers, and is sometimes placed in gardens for good plant growth.

mugglestone/ tiger iron

Mugglestone is a rock consisting of layers of Red Jasper and Hematite, and as such, all properties for either of these stones can apply to Mugglestone as well.

Tiger Iron is a stone that consists of brilliantly contrasting bands of Hematite, Red Jasper, and Tiger Eye. Tiger Iron has a golden chatoyant glow (cat's eye effect) that Mugglestone lacks.

Crystal healers use Mugglestone to help recover from anesthesia and surgery, and to reduce bleeding. It is thought to alleviate pain.

Mugglestone is thought to be useful in calming quarrels between friends or lovers. It helps to alleviate anger and stress, particularly in conflicts with other people.

Mugglestone is also thought to be useful in encouraging understanding, compassion, kindness, and empathy.

Mugglestone is thought to be a good stone for people who tend to take on the feelings and emotions of other people. It helps to keep the wearer from becoming too overwhelmed with emotion.

Tiger Iron is a good stone to help with the creative process. Artists, actors, writers, musicians, and dancers can all benefit from this stone. It is also helpful in applying creative solutions to everyday problems, so it is a good stone for people who have to think on their feet, such as teachers, businesspeople, and those in first aid.

Tiger Iron is helpful to those who are hypersensitive to noise or environmental toxins.

Muscovite Quartzite

Muscovite Quartzite is a Quartzite rock with inclusions of Muscovite Mica flakes. The reddish color of the flakes comes from iron deposits in the Muscovite, and the flakes give the entire stone a reddish sheen under the translucent Quartz.

Quartzite is a metamorphic rock that is formed when pure Quartz Sandstone is exposed to high temperatures and pressure. It has compact, well-formed grains of clear to cloudy Quartz. It is very strong, and is a raw material in the glass and ceramic industries. Quartzite is often dyed bright colors to make tumbled stones, but Muscovite Quartzite gets its color from the Muscovite in it.

Muscovite is a dark version of the mineral called Mica. Mica is best known for its strong one-sided cleavage, which makes it split into very thin layers. The layers can get so thin as to be transparent, and thus Muscovite was once used as a window glass substitute.

Muscovite is very soft (about 2-3 on the Mohs scale), but it is surprisingly durable, tough, and flexible.

Muscovite gets its name from the Muscovy state in Russia, where the mineral was used as a glass substitute in the 14th century.

Muscovite used to be known as isinglass, and it was commonly used on furnaces, to allow people to look through the furnace as they would with tempered glass today.

Today, Muscovite is ground and used to give wallpaper a shiny luster. It is also used in eye makeup and glitter. The rock Muscovite Quartzite, however, has very limited uses. Perhaps its beauty and durability will eventually make it a nice stone for cabochons and beads.

Myrrh

Myrrh is not a gemstone. It is the resinous sap from a desert tree called the dindin. The sap forms in small beads and has a very sharp, strong smell.

Myrrh historically has had an association with death. It has been used since ancient times in embalming. The ancient Egyptians used Myrrh to preserve mummies. Myrrh incense was burned at funerals.

Today, Myrrh is used in any application that warrants fragrance, such as candles, soaps, and incense. It is also found in several commercial mouthwashes, toothpastes, and healing salves.

While it is usually known for its smell, Myrrh can also be used as a flavoring for candy and baked goods.

Myrrh has a large number of healing uses. It is an astringent and antiseptic, and can be used to cure acne, boils, and other inflammations of the skin. It is also widely used to help heal infections of the mouth, such as gingivitis, ulcers, and bad breath. It is highly useful in applications of the throat, and is thought to help heal pharyngitis, bronchitis, sinusitis, asthma, and sore throat. It can be used as an expectorant (cough syrup).

Myrrh makes a helpful childbirth aid, making contractions more efficient, and relieving pain.

The word "Myrrh" comes from the Arabic *murr*, meaning "bitter."

The ancients found this resin to be so valuable, the Bible tells of it being given as a gift, along with Frankincense (another aromatic resin) and Gold, to the baby Jesus.

Modern science has recently discovered that Myrrh may be a cure for cancer. Laboratory studies show that a compound in Myrrh was effective in killing breast cancer cells. More research on this is currently in the works.

Nevada Lapis

Also known as "Lapis Nevada," this gemstone is not in any way related to real Lapis Lazuli. It is a rock that consists of eleven or more different minerals, which differ from specimen to specimen. The mineral composition determines the color of the stone. Most Nevada Lapis is of a pinkish color (which is the mineral thulite), sometimes with green, cream, blue, or white mottling.

"Nevada Lapis" and "Lapis Nevada" are the commercial names for what is more commonly known as "Nevada Stone." Its scientific name is "Thulite-Diopside Skarn." It was first discovered in the 1950s, and occurs exclusively in the Four Clovers Mine in western Nevada.

Nevada Lapis is fairly stable, both in color and composition. It can be handled roughly, and will not fade in sunlight.

Crystal healers use this stone to promote creativity and knowledge. It is said to promote courage, and is known as a protective stone. It helps balance and strengthen emotions.

Sometimes referred to as "Peace Jasper" and "Peace Stone," Nevada Lapis brings a calm, peaceful feeling to the wearer. Its pastel colors are very soothing both to the eyes and to the mind.

While the general composition of this stone is consistent, its secondary minerals vary enough that different individual stones need different cutting and polishing techniques, based on their hardness and chemical composition. Some Nevada Lapis is better suited for cabochons for jewelry, while other stones are best for carving.

In recent years, Nevada Lapis has been increasingly hard to find, and its future is unknown. Its current mine has been closed, and there are no other mines known in the world.

Obsidian

Obsidian is natural volcanic glass. It is formed when rhyolite lava from a volcanic eruption comes in contact with water, thus cooling it very fast. It cools so fast that crystals do not have time to grow, and that's why Obsidian breaks randomly, like glass, rather than along crystal lines.

Obsidian is usually black, but the presence of other minerals during cooling sometimes gives it various dark colors. Hematite in the lava makes it reddish-brown. Crystals of cristobalite in the lava result in the white specks found in Snowflake Obsidian. Bubbles of air trapped between layers of cooling lava form the unique deep stripes of Rainbow Obsidian.

When two colors of lava flow together, they occasionally mix with each other, but are so viscous that the colors do not mix; they appear side by side. This is how we get certain streaked Obsidians, such as Mahogany Obsidian.

When lava is expelled from a volcano and cools in the air before reaching the ground, it forms little balls called "Obsidian Bombs." The Obsidian Bombs found on the San Carlos Apache Range in Arizona are called "Apache Tears," and were once believed to be the tears of Apache women mourning warriors who fell to their deaths off of cliffs.

Obsidian got its name from Obsius, the Roman explorer who first discovered the stone in Ethiopia.

Metaphysical healers use Obsidian to gain clear insight into problems.

Being glass, Obsidian can be very sharp when broken. Its glassy texture makes it very easy to be knapped; therefore, it was a favorite material with which to make arrowheads.

Obsidian jewelry was found in King Tut's tomb.

In ancient times, Obsidian was used for mirrors.

Ocean Jasper

Ocean Jasper is a very rare stone, mined in one place in the world. It is found off the coast of the remote area of Marovato, Madagascar, and the deposit is so close to the ocean that it can only be mined at low tide.

Ocean Jasper is known for its beautiful orbicular designs, which are best seen in large cabochons and spheres. The patterns are spherical inclusions floating in solid Jasper, and the little circles are often heavily banded in many colors. The deposit is a massive Rhyolite (a volcanic stone) flow that was silicified: as the hot liquid magma cooled, the silica precipitated out of the magma, forming small balls.

Ocean Jasper got its name both from its proximity to the ocean, and because the patterns look like foaming bubbles.

Ocean Jasper is mined by local Malagasy natives, who gather it at low tide, take it ashore, then place it in a boat to be taken to the nearest area with a road, as Marovato is very remote and has no access other than by boat.

This is a very new stone, discovered a few years ago after a very long search. Long ago, a local fisherman told of the beautiful rock formation he had seen on the beach, and people had been looking for it ever since. Its discovery at the turn of the millennium, as well as the fact that it is found under the ocean, have caused people to nickname it the "Atlantis Stone."

Ocean Jasper is also known as "Orbicular Jasper" and "Moon Jewel Jasper," even though it is technically an agate. Madagascar natives call it "Snake Agate."

Crystal healers use Ocean Jasper to promote patience and help in meditation. It helps to bring peace of mind. The circular patterns suggest that all of nature is interconnected, and that nature is cyclical.

Physically, Ocean Jasper is said to aid in problems of digestion, and to help remove toxins from the body.

Onyx

Onyx is part of the Quartz family of stones called Chalcedony. It is a form of Agate.

Onyx occurs in several colors, but this term usually refers to the darker Chalcedonies. Thus, most Onyx will be black, brown, or gray. Onyx with white streaks is called Sardonyx. Colored Onyx is always dyed. Even most Black Onyx on the market is dyed, to achieve a more uniform color.

The word "Onyx" comes from the Greek *onux*, meaning "fingernail." In Greek myth, Eros (the god of love) clipped the fingernails of Aphrodite (the goddess of love) while she was asleep. Her fingernails fell to the earth, where the Fates turned them into the Chalcedony the Greeks called Onyx.

Crystal healers use Onyx to help with energy problems within the body. The stone is said to be able to banish excess or unwanted energy from the body, while holding in the necessary energy and assisting in challenges caused by a drain in energy. It is good for letting go of stress and banishing negativity.

Onyx is said to help soothe worries and fears, and is often used for worry stones and rosaries.

Black Onyx helps to overcome grief, banish old habits, and encourage happiness and good fortune.

Black Onyx, for all its good properties, is also said to bring on nightmares, depression, and arguments.

White onyx, on the other hand, is said to soothe quarreling couples. The stripes in Sardonyx are thought to correspond with male and female energy, so Sardonyx is often used in a romantic context.

Sardonyx was used in ancient Rome as a seal, since wax would never stick to it.

opal

Opals consist of silica spheres and water packed in layers. (About 10% of the stone is water.) The refraction of light between the layers causes the light to break into spectral colors. The size of the silica spheres determines the colors seen—larger spheres show all colors, while smaller spheres show the blue-violet range.

High quality Opal is more valuable than many Diamonds, with the best specimens going for up to $20,000 a carat.

The word "Opal" comes from the Latin *opalus*, which means "to see a change of color."

Opal is said to amplify personal characteristics (for better or worse), to bring happy dreams and good changes. Opals are associated with change in general. Because it is thought to bring about change, it is sometimes considered an unlucky stone, as some people might not be prepared for the changes. Opal is also considered unlucky because it tends to break easily.

Opal needs to be worn often to retain its beauty. Because it contains a lot of water, it is always in danger of drying out. Wearing the stone next to the body keeps it humid. It should be protected from high temperatures and from chemicals (such as household cleaners) which may be drying to it. Opals that dry out are prone to cracking and color loss.

More than 90% of today's gem-quality Opals come from Australia.

Anthropologist Louis Leakey discovered Opal artifacts in a cave in Kenya that were reportedly 6,000 years old.

The ancient Arabs believed Opals fell from the sky during lightning flashes, which gave them their color. Women in the Middle Ages wore Opals to protect the color of blonde hair. Opals were used in magic rites to promote invisibility.

opalite

Several of the stones in this book are actually glass. Opalite is the only *plastic* stone. Most Opalite is made of a plastic resin, designed to look like Rainbow Moonstone by fusing resins with metals to form the distinctive opalescence. Some Opalite is made from glass with the same optical properties. It has a milky color with a distinct blue fire tone. Held against a white background, it has a yellowish tinge.

While glass stones like Goldstone or Cherry Quartz carry the physical (and sometimes health) properties of natural glasses such as pure Quartz or Obsidian, the plastic variety of Opalite has no known healing properties.

Opalite is also known as "Moonstone Quartz," "Opalized Quartz," "Gerisol," "Lunar Quartz," and "Sea Opal."

Opalite originated in Hong Kong, and was introduced to the gem trade in 1988.

The plastic version of Opalite is very lightweight (good for earrings), and soft, so it doesn't make fine jewelry, but it is good for costume jewelry.

There is another, genuine, stone called Moss Opalite, which is a type of Opal. There is also the purple Tiffany Stone, a form of Opalite from Utah, that is a genuine gemstone.

While "Opalite" can refer to actual gems, or an inferior form of real Opal, most jewelry and carvings sold as "Opalite" are actually this man-made stone. It is very inexpensive, so any artifact called Opalite that seems pricey should be examined with caution. Some dealers try to pass off Opalite as a real stone.

peridot

Peridot is the gem variety of the stone Olivine. While Olivine is an abundant mineral, gem-quality Peridots are rare.

Crystal healers use Peridot as a good all-around producer of positive energy and healing. The gem heals bruised egos by helping to lessen feelings of anger and jealousy. It used to be used as protection from nightmares, and was thought to intensify the effects of a drug when the medicine was taken from a cup made of Peridot. It is given as a symbol of fame, dignity, and protection.

Peridot gets its name from the French word *peritot*, meaning "unclear." The frequent inclusions in larger stones was the inspiration for this name. Due to the French origins of its name, many people incorrectly pronounce it as "pear-i-doe," but its true pronunciation is "pear-i-dot."

Approximately 90% of Peridot is mined in Arizona.

Peridot is occasionally found in meteorites.

Peridot is sometimes confused with other stones, including Tourmaline and Emerald, and is called "Evening Emerald" by jewelers, due to the fact that its green is intensified under artificial light. In ancient times it was called "Chrysolite," and references have been made to this in the Bible.

The ancient Egyptians called Peridot the "Gem of the Sun." It was a favorite gemstone of Cleopatra, and the "Emeralds" she wore are now thought to have been Peridots. The Egyptians used large chunks of Peridot to carve small drinking cups for their religious rituals.

Small pieces of Peridot found in Hawaii are said to be the tears of volcanic goddess Pele.

Pirates used to wear Peridot as protection, by stringing beads onto donkey hairs and wearing them on their left arms.

Peruvian Opal

Peruvian Opals, also called "Blue Andean Opals," are a type of Opal that occurs in the Andes Mountains near San Patrico, Peru. Consisting of water trapped among Quartz spheres, they are considered "true" Opals. Because the internal crystalline structure is random, Peruvian Opals display none of the color play across the surface of the stone that most other Opals display. Peruvian Opals typically are comprised of 3% to 14% pure water, though some specimens have been found containing up to 20% water.

Peruvian Opals were valued by the ancient Incas, but they are somewhat rare today, and are rapidly rising in price.

Peruvian Opals are the color of the Caribbean Sea. They can be translucent or opaque, and can be a pure blue, or feathered with black impurities. While the natural color is usually beautiful enough to sell without color enhancement, lighter Opals are dyed a darker blue, or treated with smoke or plastic to achieve a deeper color. When purchasing Peruvian Opal, check for color variations within one stone. A real Peruvian Opal will have an even color throughout, whereas a dyed or treated stone may show uneven shades of blue.

Crystal healers use Peruvian Opals in stress reduction, to reduce fatigue, to aid in weight loss, to reduce dizziness and insomnia, and to aid in metabolism balance issues.

This stone is also sometimes sold under the name "Peacock Opal."

Known as the "Stone of Courage and Ingenuity," Peruvian Opal helps to improve self-esteem. It aids in creativity, and helps the wearer to find ways of communicating and speaking out on touchy issues with grace. It helps the wearer to find a confidante.

Being a "wet" stone, Peruvian Opal needs to be protected from heat, sunlight, and chemicals. Extremely high heat can cause it to shatter.

picasso marble

While this stone is also sometimes called "Picasso Jasper," it is not truly a Jasper. It is a Marble—a metamorphosed limestone rock. In prehistoric times, hot magma (lava) spread over limestone flats, seeping into the tiny fissures caused by heat and pressure. The magma cooled in these cracks, to bring us the dark streaks in light rock that is Picasso Marble.

The stone got its name because some people thought it resembled the wild patterns of a Picasso painting.

While this stone is fairly new in the jewelry field (it is mainly used in building and sculpture), Zuni artisans have been using it to make fetish carvings for years.

Picasso Marble is thought to assist in all areas of transformation. It is helpful when you need to renew a lost friendship, and it aids in banishing artistic blocks. It is thought to be an excellent stone for all creative endeavors, as it brings out hidden talents and skills. Like the artist for whom it was named, Picasso marble encourages the wearer to defy boundaries and stretch the limits of creativity.

This stone is believed to be helpful in weight loss, and in promoting self-discipline and respect.

Crystal healers use Picasso Marble to enable total recall of dreams.

Physically, Picasso Marble is said to be helpful in treating viral infections. It can also aid in digestive problems.

Picasso Marble is mined in only one place in the world, and that is Utah.

Marbles in general are fairly soft, but dense. They can take scratches, blows, and high heat without too much trouble.

picture jasper

Picture Jasper, also called "Picture Rock," "Ribbon Jasper," "Landscape Jasper," or "Scenic Jasper," is a member of the Quartz family, found primarily in Oregon and Idaho. It exhibits interesting black and brown "pictures" of landscapes, trees, animals, plants, and unearthly creatures, on a tan background.

While most Picture Jasper is of a tan color with darker brown patterns, different families of Picture Jasper can show different colors. Owyhee Picture Jasper from Oregon, for example, adds grays, blues, and reds to the basic color palette.

There are two ways Picture Jasper gets its pictures. One is when iron oxide seeps into tiny cracks in sandstone, or when mud seeps into gas pockets in molten lava and solidified. The other is when tiny plants get trapped in the sand during the rock-forming process, where they fossilize forever.

Called the "Stone of Global Awareness," Picture Jasper is a major stone for earth consciousness. It promotes a caring for and protection of the earth, and enhances the relationship between humans and nature. It helps to foster harmony both among nations and among friends.

Crystal healers use Picture Jasper in problems with the eyes, and it is also used to enhance visualization; to see the "big picture." It is said to help protect against pain, particularly in childbirth.

Carved into an Arrowhead, the stone is said to attract good luck.

Some Native American tribes used Picture Jasper to bring rain, and called it the "Rain Bringer."

There is evidence of Picture Jasper having been used as weapons and tools in Ethiopia millions of years ago.

prehnite

Prehnite is a mineral that is usually *not* used in jewelry. It is used for ornamental decoration and as mineral specimens, and it is occasionally cut into faceted or cabochon jewelry, but not often. Its lovely clean light green color makes it a good stone for the few jewelry items it makes.

While Prehnite is mainly seen in pale green shades, it can also occasionally be yellow, white, gray, and colorless.

This is a highly mystical stone, loved by New Age proponents as a stone of prophesy, protection, out-of-body experiences, and communication with extraterrestrial life. Called the "Prediction Stone," it is a favorite of fortune tellers.

Crystal healers use Prehnite to stimulate energy in the body, and to aid in calming. Rubbing Prehnite right before going to sleep is thought to aid in lucid dreaming. It is thought to help the wearer back to his or her original dreams, plans, and wishes.

Physically, it is said to help the immune system and encourage healing after an illness.

The stone was named for Dutch Colonel Hendrik Von Prehn. Prehn discovered the stone in the early 18th century, and it became the first mineral to be named after a person.

The Chinese name for Prehnite is *putao yü*, which means "Grape Jade," which has become an alternate common name for the stone. Its natural crystal formation looks like a bunch of green grapes.

Prehnite can change into Garnet under metamorphosis.

Also known as "Cape Emerald," it is occasionally sold by unscrupulous sellers as Emerald.

Pyrite

We've all heard of "Fool's Gold," that bane of the Old West Gold Rush. Pyrite (also called "Iron Pyrite") is a common mineral with a brassy yellow or gray-yellow metallic color. It is heavy (though not nearly as heavy as gold), and was often mistaken for gold by frustrated miners. Unfortunately, it had no commercial value. However, the rock formations that contain Pyrite often contain Gold as well, so the experienced miner, having found Pyrite, might press on in the same locale, knowing that he might find Gold in the same area.

When Pyrite is cut into small faceted gems and used in jewelry, it is called by its commercial name "Marcasite." However, Marcasite tends to tarnish easily and actually does not make a very stable gemstone.

Pyrite was once used as a major source for sulfur.

Crystal healers use Pyrite as a general aid in mental activity. It helps improve memory, aids concentration and learning, and improves communication skills. It is thought to balance creative and intuitive impulses with practical impulses, and thus makes a very good stone to wear when dealing in business and education.

The word "Pyrite" comes from the Greek *pyro* for "fire," or *pyrites lithos*, meaning "stone which strikes fire." Ancient Greeks discovered that sparks were produced when iron was struck with a lump of Pyrite. For this reason, Pyrite was often used as a fire-starting stone similar to flint.

Some Native American groups make mirrors out of Pyrite. Gazing into the mirrors, they say they can see the future, and see into a person's soul.

Despite its relative uselessness in a commercial sense, Pyrite is still a favorite of mineral collectors. Its beautiful gold metallic luster makes it a nice stone for display. The flat nodular crystal form, called "Pyrite Suns" or "Pyrite Dollars" is a particular favorite among collectors.

QUARTZ

Thought to be the most important mineral in the world, Quartz is one of the most common compounds in the earth's crust, and one of the most useful. It is used in sandpaper, soap, ceramics, televisions and other electronics, clocks, computers, and many other common items. The regular, predictable electric pulses that Quartz emits makes it a wonderful stone for electronic applications. Two pieces of Quartz tapped against each other in a dark room will show a spark.

Quartz comes in many colors, some of which have their own gem names. Examples are purple Amethyst, yellow Citrine, green Aventurine, and red Carnelian. It also comes in pink (Rose), brown (Smoky), white (Snow), and other colors. Clear Quartz is also known as "Rock Crystal." Different gem varieties of Quartz include Onyx, Chalcedony, Agate, Jasper, Sard, Chert, and many others.

Crystal healers consider Clear Quartz the most essential of their stones. Its main applications are to clarify thoughts, purify and cleanse the body, bring self-esteem, balance chaotic emotions, increase emotional energy, and focus the mind, but its clarity enables healers to apply virtually any healing property to it.

Quartz finds importance in nearly every culture. Mythology and lore about the stone is abundant all over the world. Balls made from Quartz have been used for centuries to see into the future.

The word "Quartz" is unsure in origin, but might come from the German *quarz*, which is the word for the mineral. The word "crystal" comes from the Greek *krustallos*, meaning "ice," as the ancient Greeks thought Quartz was ice formed by the gods.

Ancient Chinese believed that Quartz was formed from the breath of a white dragon, and they considered it the epitome of perfection.

Quartz has been found on the moon.

Rainbow Calsilica

Rainbow Calsilica is a very newly-discovered gemstone. It only became available as a gemstone six or seven years ago.

This stone is used often in cabochons and occasionally in beads. It has recently been seen in Zuni fetish carvings as well.

As fake as it seems (with its sharp layers of bright blue, green, brown, tan, and yellow), it is *thought* to be a real gemstone and not man-made. However, there is a lot of controversy over its true contents. Some say it is a Mexican product created by artisans from melted tiles. Some think it might be a composite material found in Mexico from dumped mining or oil-drilling wastes combined with leached minerals. One reputable source says it is composed of Calcium and Silica; others say it is microcrystalline Calcite banded with a clay called Allophane (which has a blue/green color). While theories abound that it is man-made, most people nowadays accept that it is a real gemstone. It is said to appear in seams of color in the host rock of volcanic Rhyolite.

It is believed to be a very young gemstone, having only formed in the last 30-50 years. Its youth makes it very soft, and to be used in jewelry, it is heavily stabilized, which means it is held together by a hard clear resin so that it won't crumble. (The stabilizing process is another reason why most people believe it is a naturally occurring stone, as a man-made stone wouldn't need stabilizing.) You can see this resin in the stone, as small clear spots.

It supposedly comes from only one mine in the whole world, and that place is kept a closely-guarded secret. It is rumored to be in Mexico or northern Central America.

The first specimens to come out of the mine were mainly red and yellow, with some greens and blues. More recently discovered stones have a lot more greens and blues.

This is a rare stone, and its price reflects this. Now that it is growing more common, the price has dipped slightly, but it is still very expensive.

Red Jasper

Jasper comes in many colors, and is basically interchangeable with Onyx, Agate, and Chalcedony. Jaspers tend to be more opaque and colorful. Red Jasper has a dark red-brown color.

Red Jasper, also known as "Imperial Red Jasper," is a calming stone; it harbors nurturing and protective energies. It protects against the evils of the night, and is a good stone to wear to combat nightmares and a fear of the dark. For this reason, it is a good stone for children to carry to bed with them.

Red Jasper is known to assist in rectifying unjust circumstances. It is a stone of fairness and justice.

Crystal healers note Jasper's tendency to work slowly to help in healing. It can be worn at all times, and provides constant support, particularly in the areas of blood and circulation. It can restore a lost sense of smell, and has been used in disorders of the kidneys and bladder.

Native Americans thought that Red Jasper was symbolic of the blood of the earth. It was considered to be the best stone for connecting with the earth's energies.

In Ancient Egypt, Red Jasper was used to make amulets, particularly for the protection of women, and to help them achieve grace and beauty.

Red Jasper was once thought to protect against the bites of spiders and snakes. It was thought to be helpful in protection and rescue from danger. Being an intensely protective stone, it helps in areas of survival.

In some native American cultures, Red Jasper was believed to help bring rain.

Red Jasper is considered a lucky stone for actors.

Rhodochrosite

Rhodochrosite is a gemstone that is primarily bright pink in color but also exhibits yellow or orange tones, with bands of white. Though the bands are what makes it beautiful, the general rule is the less white seen in the stone, the higher the quality.

Rhodochrosite is also sometimes called "Raspberry Spar."

Rhodochrosite is considered to be the most powerful love stone. It is said to attract the perfect love, and to enable the wearer to achieve self-love, tolerance, forgiveness, and friendship.

Crystal healers also use Rhodochrosite to raise the energy level of the wearer, and to remove avoidance and denial.

The name "Rhodochrosite" comes from the Greek *rhodos*, meaning "rose," and *chrosis*, meaning "color."

Rhodochrosite is commonly found as stalagmites and stalactites in caves. These formations give it its distinctive white banding, and gem cutters take advantage of this by cutting cabochons to show the banding. While it is sometimes found in the form of crystals (the best source of these being the Sweet Home Mine in Colorado), they are too brittle to make good jewelry. Jewelry and carvings are made from the stone in its massive form, which is soft and easy to cut.

Rhodochrosite is mildly electromagnetic, and will sometimes pick up small pieces of paper if rubbed vigorously.

The ancient Andean Indians believed that Rhodochrosite was the blood of former rulers, which had been turned into stone. This made the stone precious to them, and they buried it with their dead. When archaeologists later found the stone in Incan tombs, they nicknamed it "Rosa del Inca," or "Inca Rose," which are used as alternate names for the gem today.

Rhodonite

Rhodonite is a unique stone that is usually pink, but also comes in yellow, brown, green, and black. It sometimes turns a brownish color due to surface oxidation. It has black crystalline veins, called dendrites, running through it to give it a spiderweb look.

Rhodonite can occasionally be found as deep dark pink transparent crystals (which are intensely sought after by collectors), but it is more often found in massive form, from which cabochons and figurines are carved.

In Russia, Rhodonite was used to make serving platters, and was often given as a gift for royal weddings. Today, Rhodonite tiles can be seen adorning the walls of Russian subways.

The presence of the black marbling once made some people call it "Pink Turquoise," though it is not in any way related to Turquoise.

Rhodonite is said to help the wearer achieve his or her greatest potential. Called the "Stone of Peace," it helps bring an inner calm and patience.

The name "Rhodonite" comes from the Greek *rhodos*, meaning "rose."

Sometimes know as the "Singer's Stone," Rhodonite is said to improve sound sensitivity. In healing it is used to improve hearing loss and help inner ear problems.

Rhodonite is said to strengthen the immune system, and to strengthen the heart. It is thought to be a good stone to have nearby while recovering from a heart attack.

Another nickname for Rhodonite is the "Stone of Love." It is said to activate love energies, to attract the perfect love, and to help the wearer to love himself or herself. It also helps to foster love among all mankind, and between tribes, cultures, and nations.

Rhyolite

Rhyolite (also spelled Ryolite) is a fine-grained igneous volcanic rock with the exact same chemical composition as common Granite. It is basically a Granite that cooled too fast to form crystals.

Rhyolite is actually the geologic name for this rock, which is often dull-colored. When the brighter green forms of Rhyolite are used as a gemstone, it is known as "Rainforest Jasper" or "Australian Rainforest Jasper." The gemstones known as "Orbicular Jasper" (light background with colorful orbs) and "Leopardskin Jasper" (orange and black variety) are also forms of Rhyolite, as is the stone called "Pumice," which is used as a household cleanser and beauty aid.

Gem-quality Rhyolites have circular "eyes" called phenocysts. The colors of the phenocysts correspond with the minerals in them. White phenocysts are from Quartz, pink are Feldspar, red are Iron.

Rhyolite is said to help all forms of communication, including listening. It pushes you forward to meet your goals and dreams, and helps you to face the changes that can occur when life goals are met.

Rhyolite is sometimes worn to rejuvenate physical beauty, and to strengthen the permanency of love relationships.

Rhyolite is found in areas of volcanic activity. It is most commonly found in Australia, but varieties of it are found in volcanically active areas all over the world.

Rainforest Jasper is a good stone for connecting to the earth. It is used in earth healing rituals, and helps to forge a connection between man and nature.

Rhyolite composes the majority of the island of Komodo and surrounding islands, where the famous Komodo Dragons live.

Riverstone

Riverstone is a Jasper-related stone that typically comes in a variety of pastel colors, including pink, beige, blue, cream, and tan.

While stones that are tumbled and polished to a sandy finish by the movement of river waters are called "River Stones," the gem Riverstone is an entirely different type of stone, and it refers to a stone comprised of a specific mineral composition (as opposed to being a random stone that happens to be river-tumbled).

In addition to the light-colored Jasper called Riverstone, there is also a luminous reddish stone called "Patuxent River Stone" or "Patuxent River Agate," which is the state gemstone of Maryland, and which is commonly found in commercial gravel mixes.

Since this is a newly-discovered stone, very little information is available on it. Crystal healers haven't yet examined the full range of possible healing properties. However, even at this early stage, Riverstone is thought to be helpful in alleviating anxiety. It is a good stone when change is necessary; it promotes speedy changes and helps the wearer move swiftly through change. It is also thought helpful in reducing boredom and in breaking out of old patterns and ruts.

Because of its ability to speed change, Riverstone is a good stone for women who are experiencing delays in labor and childbirth.

Riverstone is thought to help amplify the healing properties of other stones, so it is a good companion stone in crystal healing. It also speeds up the healing properties of other stones, so when used in conjunction of other stones, the healing brought on by the other stones takes place more quickly.

Physically, Riverstone is thought to speed up metabolism, which can be helpful to those trying to lose some weight. It helps to regenerate the body and promote physical energy. It is also thought to help with problems of the teeth and bones, and with calcium deficiencies.

Rose Quartz

Rose Quartz is unusual in several different ways. Its natural pink color is rather rare among minerals, and is caused by impurities of iron, titanium, and colloidal gold. It is also odd in that it is the only type of Quartz that is not commonly found in crystal form. Rose Quartz usually comes in massive formations that are translucent to opaque; transparent crystals (called "Pink Quartz") are very rare.

Occasionally, Rose Quartz contains Rutile fibers which give it asterism—a shimmering star effect—when cut into cabochons.

The pink color of Rose Quartz is photosensitive and can fade in sunlight. It is commonly dyed to enhance the pink color.

Being pink, Rose Quartz has historically been a symbol of love and beauty. In Greek mythology, the god Eros brought this stone to earth to help arouse passion in lovers and potential lovers. It is known as the "Love Stone" or the "Heart Stone," and is said to enhance the wearer's sensitivity to beauty, art, and music. It enhances the capacity to love, and is a very good stone to give at the beginning of a relationship.

Crystal healers use Rose Quartz in all applications of the heart. It is also used to calm and balance emotions, to increase self-esteem and self-confidence, to reduce depression, to assist in weight loss, to clear the complexion and to protect against wrinkles. Its calming, soothing nature makes it a good stone to put near a fussy baby's sleeping area.

Rose Quartz is the gemstone for the second wedding anniversary, believed to strengthen the love of young married couples. It is said that keeping a Rose Quartz under your pillow will bring more sexual fire to the marriage.

Rarely seen as a faceted gem due to its opacity, Rose Quartz is most commonly seen as beads and cabochons in jewelry, and in carvings of all types, both utilitarian and decorative.

Ruby

Ruby is a gemstone of the Corundum family, which also includes Sapphire. Pure Corundum is colorless, and it is only the impurities that make it a new gemstone. All colors of Corundum are called Sapphire, with the exception of the red version, Ruby. The red color comes from impurities of chrome.

Ruby is considered the Queen of all gems. Gem-quality Rubies are so rare that it is now the most sought after and expensive gem in the world; more prized than Diamonds.

Corundum has a Mohs scale hardness of 9, which makes Diamond the only gem that is harder. Thus, Rubies are extremely durable and can be worn in any form of jewelry without special precautions.

The word "Ruby" comes from the Latin *rubeus*, which simply means "red."

Hindus consider Ruby a sacred stone, worthy of offering to their god Krishna. The Sanskrit word for Ruby is *ratnaraj*, which means "King of Gemstones."

While most Rubies are faceted, some Rubies contain Rutile that gives them a star effect (asterism) when cut into cabochons. These "Star Rubies" are highly prized. The largest Ruby in the world, at 2,475 carats, is a star Ruby.

Crystal healers use Rubies in all applications of the heart, and to purify blood. It is said to amplify energies, both good and bad. Ruby opens the heart to love and gives the wearer the capability to soothe lovers' quarrels. It is worn to ward off misfortune and ill health, and to promote longevity and stable finances.

It was once thought that dreaming of Rubies would predict coming success in business, money, and love. Rubies were once thought to grow darker when the wearer was in danger or when an illness was coming. If worn on the left side of the body, Rubies were thought to give the ability to live in peace among enemies.

Ruby Zoisite

Zoisite is a mineral that can be colorless, white, pink, blue, purple, yellow, brown, green, or red. The Zoisite in Ruby Zoisite tends to be green with black streaks. It is a natural matrix for the growth of Ruby, and thus the two are often found in the same stone.

Ruby Zoisite is Zoisite is found with bits of Ruby in it, and it is also known as "Anyolite."

Ruby Zoisite is found almost exclusively in Tanzania, along with the blue variety of Zoisite, the gemstone Tanzanite.

Zoisite was named after Seigmund Zois, the Austrian natural scientist who discovered it. The word "Anyolite" comes from a Masai word meaning "green."

Ruby Zoisite is thought to turn negativity into positive energy. It facilitates the advancement of all mental talents and abilities, amplifies the energy field in the body, and helps to relieve lethargy and boost mental and physical energy. It helps to rid the wearer of laziness and idleness.

This stone promotes a sense of one's own individuality, and helps the wearer to avoid conforming to other people's ideas and ways.

The contrasting colors of this stone (red and green), make it an excellent stone for balancing male/female energies.

Physically, Ruby Zoisite is thought to strengthen the heart. It stimulates fertility, and helps in diseases of the testicles and ovaries.

This stone is fairly unique in that it has two different hardnesses. The Rubies in it are extremely hard, while the green Zoisite is much softer. Because of the inconsistencies in hardness, Ruby Zoisite is a very difficult stone to cut.

Rutilated Quartz

Normally, the fewer inclusions to be found in a gem, the higher the value. Rutilated Quartz, however, is valuable *because* of its inclusions. Also called "Rutile Quartz," this is a rock made from crystalline Quartz embedded with impurities of titanium dioxide, which is called "Rutile." Rutile grows in long, thin crystals, and within the Quartz, the crystals look like tiny reddish or golden needles.

The Rutile inclusions amplify the metaphysical properties of the Quartz. Thus, all of Quartz's properties apply to Rutilated Quartz, only more so.

This stone is also known as "Venus's Hair Stone," "Cupid's Darts," and "Fleches d'amour" (which is French for "Arrows of Love"). A less common name for this gem is "Sagenite."

Rutile has a hardness of 6 on the Mohs Scale, whereas Quartz has a hardness of 7. Due to the differences in hardness, it is difficult to cut Rutilated Quartz in such a way that it does not exhibit pitting across the surface of the bead, faceted gem, or cabochon. The best gemstones are those in which the Rutile fibers are completely surrounded by Quartz, so that none hit the surface of the stone.

Rutilated Quartz is a highly energetic stone. It helps energy move freely through the body, and assists mental focus. It is helpful in uncovering the causes of mental hang-ups.

This stone helps to diminish fears, aid in decision-making, ease loneliness, enhance self-reliance, and reduce guilt. It helps the wearer to improve his or her creativity, to find creative potential, and concentrate on personal skills and talents.

The word "Rutile" comes from the Latin *rutilus*, which means "red."

The best-known Rutilated Quartz is crystal clear. However, it can also come in milky-white and Smoky Quartz.

sandalwood

Not truly a "stone," this herb is occasionally made into jewelry, so it is being featured in this book. Sandalwood is a small evergreen tree with yellow to maroon flowers. It lives as a parasite for the first seven or eight years of its life. It is native to India, but will also grow in similar climates.

The essential oil comes from the heartwood deep in the mature tree. It takes about 50-60 years for a tree to be ready for harvest. The tree is harvested by uprooting rather than cutting, because the most fragrant oils are found in the roots. The bark and the flowers have no scent at all.

At the time of harvest, careful records are kept of every scrap of wood, including the smallest chips and sawdust. (Today, Australian plantations are growing new groves, but it will be another half-century before these trees are ready for commercial harvest.) Despite all Sandalwood trees in India belonging to the government (including those on private property, for which the owner is entitled to 75% of the tree's value upon harvest), illegal poaching of trees happens regularly. Currently, there is only one Sandalwood producing factory in India that is permitted to export the oil and wood, but this doesn't stop smugglers from exporting it.

Sandalwood is thought to have been in use for over 4,000 years. Its fragrance is warm and sensual, unlike any other in the world. It is used extensively in religious ceremonies.

Used externally, Sandalwood is thought to be good for skin disorders of all types. It moisturizes aging skin, and is used as an astringent. Internally, Sandalwood aids in disorders of the stomach and digestive system. It is useful in reducing fever, especially fever associated with sunstroke.

Aromatherapists use the sensual fragrance of Sandalwood in a large variety of situations. It can lift depression, help one to breathe more clearly, and reduce stress. It is used as a sleep aid and an aphrodisiac.

The Hindi word for the Sandalwood tree is *chandan*.

Sapphire

The mineral Corundum is clear and colorless. When it is colored red or deep pink from chemical impurities, it is called Ruby. When it is any other color, it is called Sapphire.

While Sapphires can come in virtually any color other than red, the cornflower blue type is the most highly valued. The blue color comes from impurities of iron and titanium. Sapphires usually show pleochroism, which means that they look different colors when viewed in different lights. Darker-colored Sapphires ore sometimes considered to be "male" stones, while the lighter colors are "female" stones.

The most sought-after Sapphire is the "Padparadscha," which is a bright salmon color.

The name Sapphire comes from the Latin *sapphirus*, which means "blue."

Crystals healers use Sapphire to treat colic, disorders of the blood and skin, depression, poor hearing, and nosebleeds. Sapphire is thought to promote generosity, good manners, wisdom, noble thoughts, peace between enemies, and protection against fraud and envy. It was once thought to darken with infidelity, so husbands would give them to their wives before leaving on extended journeys. It encourages truth, sincerity, and devotion, and therefore makes a good stone for an engagement ring.

Traditionally, Sapphires have been associated with royalty. Kings would wear them for protection.

Ancient Persians believed the earth rested on a Sapphire, and that its color was reflected in the sky.

Originally, the word "Sapphire" was used to describe only the blue stones. Colored Sapphires were called the stone that corresponded with their color, with "Oriental" before it. (For example, a green Sapphire was called "Oriental Emerald," a yellow Sapphire was called "Oriental Topaz," and so on.)

Serpentine

Serpentine is a combination of several minerals in differing concentrations. Thus, there are many different looks to this stone. "Antigorite" is a dark green color while "New Jade" (which is not a real Jade) is a light green. A stone with a mixture of Serpentine and Chrysotile has the trade name of "Infinite Stone."

The mineral Serpentine, which colors the rocks used in gems, is rather common. Most rocks that contain some green usually have Serpentine in them.

The name Serpentine comes from the fact that it often has a scaly appearance like snakeskin.

Having various hardnesses and being a fairly soft stone overall, Serpentine is more suited for decorative items such as figurine carvings than for jewelry.

Serpentine's green color and waxy, smooth surface are reminiscent of Jade, and it is often sold as Jade by unscrupulous sellers.

Crystals healers use Serpentine to help reduce swelling, and to ease the symptoms of asthma. It is said to improve the milk production when worn by nursing mothers.

Serpentine is thought to bring happiness, success, friendliness, and independence. It brings respect for the elderly and the wisdom they are able to provide. It is thought to be particularly protective of fathers, and so it makes a good gift for Father's Day.

Serpentine was highly prized by the Aztecs, who valued all green stones.

Long ago, Italian witches (*streghe*) believed that carrying small pebbles of Serpentine would protect them from the bites of animals such as snakes. If the person had already been bitten, the stone was thought to be able to pull the toxins out.

Silver

Silver occurs on its own in nature, or as an ore in host rocks. It is slightly harder than gold, and fairly rare and expensive. It is slightly less valuable than Gold, however, due to its tendency to tarnish when in the presence of ozone, hydrogen sulfide, or sulfur.

Silver has been used since ancient times. There is evidence that humans learned how to separate Silver from lead as early as 3,000 B.C. In 700 B.C., Silver was first used to make coins. The words for "Silver" and "money" are identical in at least 14 languages.

Besides coins, Silver has many other uses. Its most predominant use is in photography. It is also commonly used as a precious metal for jewelry, as an aid in dentistry, and in electronics, among many other things.

Silver is the best conductor of heat and electricity known. However, due to the high cost of Silver, the less-expensive Copper is more readily used in electronics.

The word "Silver" comes from the Anglo-Saxon *seolfor*. Its chemical symbol is Ag, from *argentum,* the Latin word for Silver.

Commercial fine Silver is 99.9% pure Silver. Sterling Silver, which is widely used in jewelry, is 92.5% pure Silver.

Silver compounds can be absorbed by the skin and tissues, resulting in a blue or blackish pigmentation called argiria.

Silver has historically been associated with the moon. In Greek mythology, Artemis, the goddess of the moon, carried silver arrows that caused swift, painless death to those who were shot with them.

In the early 1900s, people would put Silver dollars in milk jugs, believing it would keep the milk fresh.

Smoky Quartz

Smoky Quartz (also spelled "Smokey Quartz") is named for its color, which is light tan or brown or gray. The color is thought to be the result of chemical impurities in the Quartz, but it is also widely believed that the color is due to small amounts of radiation. Natural Smoky Quartz usually occurs in rocks with a small but persistent amount of radioactivity, and the stone is sometimes artificially created by irradiating Clear Quartz to darken it.

The color of Smoky Quartz is rather rare in the mineral world, as there are only a few other naturally dark brown stones.

There are three famous types of Smoky Quartz:
1. Cairngoran—A light color that originates in Scotland.
2. Morion—A dark brown or black variety.
3. Coon Tail—A dark and light striped type.

Smoky Quartz is thought to be a warm, soothing stone, bringing calm and comfort to the wearer. It helps to heal grief, and promotes creativity.

Smoky Quartz is said to be beneficial to groups, fostering cooperation.

The energy in Smoky Quartz is more subtle than the energy of Clear Quartz, so the effects of the stone happen gradually.

This stone carries the anchoring energies of the earth, and is helpful in creating a deeper appreciation of the earth, and a greater concern for the environment.

Smoky Quartz is occasionally sold as "Smoky Topaz," which is a misnomer, as it has no relation to Topaz.

Smoky Quartz is the national gem of Scotland (their national scepter has a large Smoky Quartz on top), and has been considered sacred there since the time of the Druids.

Snowflake Obsidian

Snowflake Obsidian (also known as "Flowering Obsidian" or "Spherulitic Obsidian") is formed when crystals (called spherulites) get caught in the cooling lava and form the white-gray cristobalite that makes the "snowflake" look.

While most Obsidian is chemically glass, Snowflake Obsidian is the one member of the Obsidian family that is actually a rock. It is similar in composition to Granite (which is slow-cooling and forms large crystals), and Rhyolite (which is more quickly cooling and forms much smaller crystals). Snowflake Obsidian cools faster than either of these, so it obtains the glassy texture typical of Obsidian.

Like a volcano brings lava to the surface, Snowflake Obsidian brings emotions to the surface of the brain. Both positive and negative things are brought to the surface, where the user is able to examine and deal with them.

Crystal healers use Snowflake Obsidian to transform negative energy into positive energy, so it is a good stone to use at the beginning of a healing session.

Snowflake Obsidian is said to lessen fear, bad dreams, anxiety, and grief. It helps the wearer become more aware of danger in order to stay safe.

Physically, Snowflake Obsidian is used to treat muscle cramps, and it is a good detoxification stone.

Old lore says that if you keep a piece of Snowflake Obsidian with your money, it will prevent the money from running out.

Snowflake Obsidian in known to be one of the first stones to be used for scrying (fortunetelling).

sodalite

Known as "Poor Man's Lapis" due to its similarity in color, Sodalite is actually a component of Lapis, but is much less expensive. It comes in all colors of blue, with white Calcite veins. It also less commonly comes in gray, white, pink, green, yellow, and other pale colors.

Alternate (though not common) names for Sodalite are "Canadian Blue Stone" (as one of its main deposits is found in Ontario), and "Bluestone."

Sodalite is thought to help change the wearer's thoughts about himself or herself. It helps the wearer to be less self-critical, be more introspective, and to analyze goals.

Sodalite is also believed to be a good stone for groups working together, as it forges a sense of solidarity and trust. It aids in communication and in writing.

Crystal healers use Sodalite to draw out infection, to reduce fever, and to cure insomnia. It shields the wearer from negative energy, and blocks radiation (and thus it is a good stone to keep by a television or computer). It stills the mind, calms the spirit, and relaxes the body, and is a good stone to use in meditation.

The name "Sodalite" refers to its high Sodium content.

While Sodalite is usually found in opaque massive form, transparent crystals have occasionally been found.

Most Sodalite will fluoresce (glow under a black light), and a certain type of Sodalite called "Hackmanite" displays the unusual characteristic of tenebrescence, which means it changes color when exposed to light (and often back again when put in the dark).

The stone is relatively rare, with only three known mines in the world.

spinel

This stone has been known in the gem world for only about 150 years, but has been around much longer, as a Ruby impostor. Due to its close chemical composition to Ruby, Spinel has throughout the ages been mistaken for Ruby. Two very large "Rubies" in the British Crown Jewels are actually Spinels.

Spinel comes in a variety of colors, but red and pink are the most popular. Even though it is not Ruby, it still is a well sought-after gemstone, and its value is rapidly approaching that of Ruby.

Fine red Spinels can command $500-1,500 per carat.

The crystals occur in well-formed octahedrons.

The name "Spinel" is unclear in its origins. Some think it stems from a word that means "spark" or "point." It may come from the Latin *spina*, meaning "thorn."

In Medieval times, Spinels were called "Balas Rubies" or "Lal."

Some Spinel, called "Lodestone," has magnetic properties, and was once used by mariners to magnetize their compasses so that they could be guided at sea.

Crystal healers promote Spinel for happy marriages. The stone is thought to set aside the ego to enable one to become devoted to someone else. It helps to encourage great passion, and provides longevity. All three of these are helpful for a happy marriage.

Even though it is a relatively new discovery, there is still a lot of lore surrounding this stone. A loss of luster is said to warn of danger. It was once thought to protect the wearer from fire. Powdered Spinel was thought to make a potion to detect when others are lying.

sugilite

Sugilite is a fairly newly-discovered mineral. It was first found in 1944 by Japanese geologist Ken-ichi Sugi, after whom the stone was named. It is found in only a few places on earth, and can be found as deep as 3,500 feet under the ground. Because of its depth and difficulty in obtaining, the stone is very expensive.

Sugilite is light gray-purple to deep violet. The darker the color, the more valued the stone. It usually occurs in massive opaque form, though a few transparent crystals have been found. This "Sugilite Gel" is extremely valuable.

Sugilite is also known as "Royal Azel," "Luvalite," and "Purple Turquoise" (though it has no relation to real Turquoise).

This stone is highly valued by crystal healers. Called the "Healer's Stone," it is believed to be a good general healing stone, and to amplify the effects of other stones. It is thought to be a wonderfully psychic stone. It stimulates the "3rd Eye," the energy center in the middle of the forehead, which helps with enhancing inner vision. It protects against "Psychic Vampirism," whereby energy is sapped from a person by others. It works as a "comfort stone" for highly sensitive people, giving them security against hostility and other negative emotions. It lessens shock and disappointment, eases severe changes in life, and enhances belief in one's self.

Sugilite helps the wearer understand the spiritual reasons for lessons or conditions realized on earth. As an example, once believed to be able to cure cancer, now Sugilite is thought to not cure it, but help the wearer figure out why he or she developed the disease, and to help find ways to treat it.

Physically, Sugilite is used to heal pain and discomfort of all kinds.

Crystal healers believe they must "cleanse" this stone periodically by placing it in full sunlight or moonlight for a whole day or night. This removes the spiritual impurities and makes it an excellent stone for meditation.

sunstone

Sunstone is a type of feldspar with a metallic glitter caused by tiny Hematite and Goethite platelets. This causes the stone to have a deep, rich shimmer called "schiller." The feldspar crystals form in molten lava and are brought to the earth's surface during volcanic eruptions.

There are essentially two types of Sunstone. One is found in India, Norway, and certain areas of the United States, and the other type is found only in Oregon. Oregon Sunstone contains small amounts of copper to give them a reddish glitter. The copper causes the schiller to form in snowflakes, stripes, or sheets; thus, every individual stone is quite unique.

Also called "Heliolite," "Red Labradorite," and "Aventurine Orthoclase," Sunstone is *not* Labradorite (although it is in the same family) or Aventurine. It is often confused with Aventurine, and with the glass stone known as Goldstone, but Goldstone is obviously different, with more uniform sparkles. Being in the same family as Labradorite, it exhibits a similar sparkle.

Oregon Natives used Sunstone for bartering. Canadian Indians used it in medicine rituals to help the spirit guides access the healing power of the sun. Ancient Greeks used it in goblets to prevent poisoning.

Sunstone is thought to dispel fear and stress. It is thought to increase vitality, originality, independence, and courage. It helps to promote a sunny outlook, and provides clarity in decisions.

Metaphysical healers use Sunstone to enhance leadership qualities in the wearer.

Ancient Greeks believed that Sunstone kept the sun on its daily journey, and they held it in such high regard that they dedicated it to the sun god, Helios.

Vikings believed Sunstone would guide them into the afterlife, and they would bury their dead with Sunstone to help guide them through Valhalla.

tanzanite

Tanzanite is a relatively new gemstone, introduced to the world in 1969 by Tiffany & Co. It was first discovered in 1967, in Tanzania (hence the name). Masai cattle herders were roaming in a place where a lightning storm had caused a recent fire. The brownish crystals that they had occasionally seen in the area were now a brilliant blue. Tanzanite, which usually occurs as a brown crystal, turns vivid blue when treated with heat; thus, nearly all gem-quality Tanzanite has been heat-treated.

Tanzanite tends to look different colors in different lights. In natural sunlight, it looks purple. By light bulbs and candlelight, it appears more brown. Fluorescent lights and overcast lighting bring out the blue. This gem also exhibits trichroic pleochroism, which means it shows three colors in one stone (bronze, purple, and blue) when viewed from different directions, giving the stone beautiful depth. Heating the stone brings out the blue color, but reduces the pleochroism, so gem-quality blue stones look "flatter" than rough crystals.

The bluest stones are the most valuable. The purple stones, looking a lot like Amethyst, are not as valuable.

Several things contribute to Tanzanite's value and thus its high cost. It is mined in exactly one location in the world—the Merelani mine located between the two historic African landmarks, Mount Kilimanjaro and the Olduvai Gorge where anthropologist Louis Leaky found his most famous early human fossils. The hand-mining process is very laborious, which adds to its cost. At one time it was abundant, but then in 1998 floods entered the mine, killing over a hundred miners. Since then, the price has remained very high.

While beautiful as a gemstone, Tanzanite is fairly soft and brittle, easily scratched and chipped, so great care must be taken to avoid heat and shock.

Being very new to the gemstone market, crystal healers have not made much claim on Tanzanite, though African lore says it is good for intense, high-strung people who need to slow down and mellow out.

tiger eye

Tiger Eye (also called "Tigereye," "Tiger's Eye," "Cat's Eye," or "Crocidolite") is a compact form of Quartz, with fibrous, parallel inclusions of crocidolite asbestos, which turn brown upon oxidization. These inclusions catch the light and give Tiger Eye the effect known as chatoyance, the shimmering effect that likens the stone to a Tiger's Eye.

Crystal healers use Tiger Eye to strengthen bones, aid digestion and cure ulcers, and in healing problems with the eye. It is said to increase night vision.

Tiger Eye is said to bring wealth, courage, strength, and joy. It is used to lift the spirits; to bring the wearer out of "the blues."

Tiger Eye is usually golden in color, but also comes in green, blue, black, red, and brown. Red Tiger Eye is called "Dragon's Eye" or "Ox Eye." Blue Tiger Eye is called "Hawk's Eye." Hawk's Eye is Tiger Eye before the blue-gray Crocidolite has oxidized. When golden and blue Tiger Eye occur together, it is called "Peitersite."

Tiger Eye is known as the "Stone of Perception." Its constantly changing appearance reminds us that life is always changing, and the stone helps the wearer to accept change and new directions in life.

Roman soldiers used Tiger Eye for protection in battle. Its eye appearance was thought to provide a safe lookout for warriors.

It was once thought that holding Tiger Eye would make it possible to see through closed doors.

All "eye" stones have been used throughout history as protective talismans. Today, Tiger Eye is still seen as protective.

It is rare to find a piece of Tiger Eye that is larger than two inches across. Because of this, figurines carved from Tiger Eye are usually small.

topaz

Topaz, called "The Stone of Love and Success in All Endeavors," comes in a variety of colors, the most common being golden-yellow and blue. Pure Topaz is colorless, but chemical impurities give it various colors. It is often pleochroic (it looks different colors when viewed from different angles), so cutters must take care to cut to the best advantage of the stone.

Some Topaz varieties include "Blue Topaz," "Orange Topaz," "Imperial Topaz" (a yellow-to-peach color), "Sherry Topaz" (yellow-brown), "Rose Topaz" (pink), and "Mystic Topaz," which is colorless Topaz that has been coated with a thin metallic layer to form iridescent colors.

There is some controversy over how this stone got its name. Some believe that the word "Topaz" comes from the Sanskrit *tapas*, meaning "fire." Others think it was named after the Topasos Island in the Red Sea, where Peridots were once erroneously called Topaz. The name was given to the original golden-yellow stone, and the same gemstone in different colors is still called Topaz, with a qualifier before it.

Some Topaz crystals can grow to massive sizes, weighing hundreds of pounds. The largest Topaz crystal found weighed 597 pounds. The world's largest faceted gemstone is a 36,854 carat Topaz.

The ancient Greeks believed that Topaz would afford the wearer invisibility in threatening situations, and also that it would change color in the presence of poison.

Topaz is used by crystal healers to treat high blood pressure, depression, insomnia, cough, asthma, and hemorrhage. It is thought to reduce anger and increase happiness. In ancient times, the stone was used to heal problems of the eye. It was soaked in wine for three days before being applied to the eyes.

The ancient Egyptians thought Topaz was colored by the sun god Ra, and they used it as a protective amulet.

tourmaline

Crystal healers love this stone. Its many colors gives it many healing applications. Among others, Tourmaline is said to strengthen the skeleton and heart, and balance moods. It is soothing, calming, uplifting. Old lore says it was worn as protection from danger. Tourmaline is said to be a favorite to encourage artistic expression—it has many colors and expresses every mood.

The gem Tourmaline is called "Elbaite," and it comes in a variety of colors. The pink variety of Tourmaline is called "Rubellite," and the green is called "Verdelite." Elbaite includes the famous "Watermelon Tourmaline," a crystal that is green on the outside and pink on the inside. When the crystal first begins to grow, it is heavy in Rubellite, and later on, the crystal faces chemical changes which turns it into Verdelite, thus completing the growth of the crystal in green.

Blue Tourmaline is called "Indicolite," and is extremely rare and sought after by collectors.

Black Tourmaline is called "Schorl." It is typically too dark to be used as a gemstone, but needles of Schorl are sometimes found in pure Quartz, making the beautiful stone "Tourmalated Quartz" or "Tourmalinated Quartz."

Tourmalines are pleochroic, so they look darker when viewed down the long axis of the crystal than when viewed down the side.

Watermelon Tourmaline is said to be a good stone for married couples: the pink balances female energy, and the green balances male energy.

The word Tourmaline comes from the Sinhalese word *tourmali*, meaning "mixed."

Tourmaline is often used in electrical applications because it creates an electrical charge when put under pressure. The electric charge is sometimes so great that you can see a tourmaline stone collecting dust and ashes. Long ago, people used Tourmalines to clean out stove pipes.

tsavorite

Tsavorite is in the Grossular family of Garnets. It is a brilliant green color.

Tsavorite was discovered in the early 1960s, in Tanzania. Since then, it has been mined in Tanzania and Kenya. It is an incredibly rare stone due to only being found in one part of the world, and due to the fact that the mines are in very inhospitable environments, where poisonous snakes, lions, and other dangerous animals dwell. (The man who first discovered this stone, Campbell Bridges, was killed by claim jumpers on his mine in 2009.) Less than 1,000 carats of Tsavorite are mined in a single year.

It is very rare to find a Tsavorite stone that is larger than one carat. Stones of two or more carats are nearly impossible to find.

Tsavorite gets its name from the Tsavo National Game Reserve in Kenya. It is also known as "Tsavolite."

Emeralds are facing great commercial competition from Tsavorite as the world's premier green gemstone. Not only is Tsavorite rarer than Emerald, but it is greener, brighter, stronger, and does not need any types of treatment (heat, oiling, dying, irradiation, etc.), whereas Emerald is a fragile stone that needs to be oiled in order to be worn as jewelry. However, jewelers are reluctant to carry Tsavorite, because the general public prefers to buy more well-known stones such as Diamond, Emerald, Ruby, etc. Plus, its rarity makes it difficult to supply, and gem dealers prefer to have a reliable source of gems.

Crystal healers use Tsavorite to help strengthen the body, promote energy, and protect from poisoning and infectious diseases. It is thought to help repel insects and heal irritated skin.

Tsavorite helps the wearer to find wisdom and the courage to guide one's own destiny. It enhances the imagination, intuition, and self-understanding. Tsavorite was once thought to protect from bad dreams when placed under a pillow.

Turquoise

Turquoise gets its characteristic sky-blue color from a large amount of copper in the stone. "Spiderwebbing," the dark-colored marbling that flows through some Turquoise, is created when iron oxide fills cracks in the stone.

Turquoise is considered a "lucky" stone in many cultures. In some Native American cultures, it is thought to bring the energy of the sky down to earth.

Turquoise helps the wearer to stay centered and "at home" in any environment. It is said to bring wisdom, understanding, trust, and kindness. It helps initiate romance, and this makes it a good stone to give to a new love.

Turquoise draws negative vibrations from the body. When placed at the feet or in a sock, the stone draws the negative energy out of the body and returns it to the ground.

Turquoise is one of the oldest gems known. It was used by Egyptians in 6000 B.C. or earlier.

The name comes from the French *turquoise*, meaning "Turkey." The best Turquoise is found mainly in Iran and Turkey.

Turquoise is used to gain wealth, to protect from environmental pollutants, and it is thought to change color when the wearer is in danger.

In the Americas, Turquoise only started to be used as jewelry in the 1880s, when a Navajo craftsman used Turquoise with coin silver to make jewelry. Prior to that, it had been used for carvings, mosaic tiles, and beads.

It is common to see saddles and bridles decorated in Turquoise. This stems from an old belief that the stone will protect horse and rider from injury. When both rider and horse wear Turquoise, the stone will absorb the injury during a fall.

Turritella

Turritella (also spelled "Turitella") is composed of fossils embedded in a limestone matrix. The fossils consist mainly of snails and crinoids.

This stone is also sometimes known commercially as "Fossil Agate," "Fossil Stone," or "Fossil Limestone." It's also sometimes called "Turritella Agate," though it is not an Agate.

This is a very soft stone, prone to chipping and shattering, and should be treated with care.

The stone Turritella is named for the mistaken belief that turritella snails are among the fossils found in the stone.

Turritella is rich in lore and metaphysical belief. Fossils traditionally are thought to extend the life of the wearer; thus Turritella is said to bring longevity. Fossils also have traditionally been used to strengthen bones and teeth.

Turritella is thought to help the body absorb nutrients from food, and to soothe an upset stomach. It is also used to reduce fatigue.

The stone is worn to help overcome fears and insecurities. It aids in easing domestic relations and group interaction. It is also thought to help soften a strong ego, or to support a weak one.

Being made from the bodies of once-alive creatures, Turritella is thought to open the wearer to nature, and to be able to communicate with the world of plants and minerals.

The hard fossils in a soft stone are thought to help the wearer combine old and new; to make transitions easier.

unakite

Unakite is a relatively new stone for jewelry. It is composed of olive-green Epidote and pink Feldspar, which give it its distinctive green-pink mottling. Because of its Epidote content, it is sometimes known as "Epidotized Granite." Some people mistakenly call Unakite "Epidote," but Epidote is actually just one component of the stone.

Crystal healers use Unakite in the treatment of the reproductive system. It is said to be a good stone for pregnant women to wear, as it facilitates a healthy pregnancy and good health for the unborn.

Unakite is associated with balancing emotions to help the wearer deal with issues from the past and help to live in the present. It removes the emotional blocks that keep the wearer from moving forward. It helps the wearer see the brighter side of things, and can lift the spirits. Some people like to keep Unakite in every room of the house, to keep the flow of joy and happy spirit even throughout the house.

Other health treatments include aid in gaining weight, and in the diagnosis of elusive illnesses. A large piece of Unakite is said to facilitate healing of anyone in poor health.

Some lore says that if you have lost an item, holding a piece of Unakite in your hand will help you find it.

Unakite is named for the Unaka Range in South Carolina, where it was first discovered.

Unakite is a hard, durable stone that is good for any type of jewelry. It does not form crystals, so it is always made into cabochons or beads. It is also used extensively for decorative objects such as figurines, eggs, spheres, and other shapes.

Unakite is said to be helpful in gardening.

Variscite

Variscite is a rare stone with a bright green color. It is often confused with Turquoise and Chrysocolla, but it is not related to either. It is a hydrated aluminum phosphate that gets its bright green color from chromium rather than from copper.

The name "Variscite" comes from *Variscia*, which is the old Latin name for Vogtland, Germany, where the stone was first found. Variscite is occasionally found in Fairfield, Utah (very near Salt Lake City), and one of its alternative names is "Utahlite."

While most Variscite is opaque, some very high quality stones can be translucent. These are set into settings with open backs, to let light shine through the stone.

Variscite is an excellent calming stone. Known as one of the true "worry stones," it eases fear, tension, anxiety, and stress. It encourages truthfulness, courage, virtue, and hope. It is a good stone to be used by anyone who works with the elderly or sick, as it promotes inner strength and the ability to deal well with the stress of caring for others.

Physically, Variscite is thought to be good in disorders of the nervous system, blood flow, and male reproductive system. It helps to aid concentration, and fights chronic fatigue.

Variscite has various hardnesses, but is a soft stone overall. Care should be used when wearing it as jewelry. Chemicals, steam, and hot water should be avoided, and it isn't a good idea to let Variscite soak for long periods of time in water of any temperature.

Variscite is often cut into figurines, and gem cutters like to use the matrix in the stone to enhance the art of the carving.

Some Variscite has inclusions of Gold in it. The Gold can be very fine, or can be in flakes large enough to see.

Wild Horse Jasper

Wild Horse is the name given to a type of Jasper that exhibits the spotted patterns of a pinto pony.

Wild Horse Jasper is a rather new stone, discovered in the 1990s near the Globe Copper Mine in the Gila Wilderness Area in southern Arizona, while mining for Magnesite. To this day, this is the only area in which it has been found.

There is another stone called "Wild Horse Picture Jasper" (note the word "picture" in the name), which is a type of Owyhee Jasper from Oregon and Idaho. The stone detailed here, however, is found only in one mine in Arizona. Its rarity is what makes it so expensive, as there is only one source for it in the world.

Also called "Wild Horse Magnesite," it is a stone made from Magnesite (magnesium carbonate) mixed with Hematite. The Magnesite provides the white background, while the Hematite makes dark brown irregular spots on the stone.

Other names for this stone include "Crazy Horse Jasper" or "Crazy Horse Magnesite," and "Appaloosa Jasper" or "Appaloosa Magnesite," as it resembles an Appaloosa pony.

It is sometimes called "Wild Horse Turquoise" or "White Turquoise," though it is not Turquoise. However, Wild Horse Jasper is found in similar deposits as Turquoise, and can sometimes contain small amounts of Turquoise in it.

A fairly hard, durable stone, Wild Horse Jasper is good for any jewelry applications, and does not require any special care. It has been popular in Western-style jewelry, paired with silver and Turquoise. It is also a good stone for carving into figurines.

Zebra Jasper

The root word for Jasper means "spotted stone" in Greek, yet Zebra Jasper is striped. It is usually a light gray or dull green with dark gray to black stripes.

This stone is also known as "Zebra Stone," "Zebra Marble," "Zebra Rock," and even "Zebra Agate" (though it is not an agate).

Mystics believe that the light stripes inspire faith, joy, and optimism, while the dark stripes inspire endurance, confidence, and strength.

Crystal healers like Zebra Jasper for its ability to ease depression and anxiety, and steady mood swings. Physically, it is said to stimulate energy and provide stamina and endurance, and therefore makes a good stone for athletes. It also helps the wearer to complete projects that require a lot of concentration and focus.

It is also thought to be good for problems of the bones, teeth, and gums. Crystal healers use it to alleviate arthritis, osteoporosis, and toothache.

Zebra Jasper's even contrast of black and white draws the mind to see both sides of any situation. It helps the wearer to find a good balance in situations where the answer is unclear.

Zebra Jasper also helps the wearer to stop considering failure in any scenario. It helps the user eliminate apathy and to take action toward positive goals, to ultimately result in success, while at the same time allowing the wearer to relax and focus on life around them, in order to understand the best course of action for success.

Its patterns being reminiscent of an animal's coat, Zebra Jasper is a good stone to help foster awareness of and communication with nature, especially with animals.

additional gemstone information

Mohs Scale of Hardness

Throughout this book are references to the Mohs scale. Some stones will be said to be a low number on the Mohs scale, which implies that they are on the soft side. Other stones will be given a high number on the scale, which means they are harder.

The Mohs scale is the official scale of mineral hardness which measures scratch resistance from one mineral to another. Created in 1812 by the German geologist Freidrich Mohs, it is the widely accepted measure of a stone's hardness.

The scale is ordinal only, which means that the degree of hardness from one number may not necessarily be the same degree of hardness from another number. For example, Diamond, at number 10 on the scale, is four times as hard as Corundum at 9, but Corundum is only two times as hard as Topaz at 8.

Where a stone falls within the scale is determined by the hardest mineral that the stone can scratch, and/or the softest mineral that can be scratched by the stone. For example, if a stone can be scratched by Topaz, but not by Quartz, it is given a hardness of 7.5. The Mohs scale is not an exact science, but it does give a general guideline for stone hardnesses.

MOHS SCALE OF HARDNESS

1. Talc
2. Gypsum
3. Calcite
4. Fluorite
5. Apatite
6. Orthoclase Feldspar
7. Quartz
8. Topaz
9. Corundum
10. Diamond

Rockhounds (rock collectors) trying to identify certain stones via the Mohs scale often carry certain items with them to help aid their scratch tests. Typical handy items that can help determine simple hardness include:

> Pencil "lead" (graphite): hardness 1.5
> Fingernail: hardness 2.5
> Copper penny: hardness 3
> Knife blade: hardness 5.5
> Glass: hardness 5.5
> Porcelain plate: hardness 7
> Sandpaper: hardness 9

Using any of these objects to test the hardness of a stone helps to place the stone on the scale, and can aid in identifying it. For example, an oily, pale stone that looks like it could be either Gypsum or Calcite can be tested by scratching it with a fingernail (hardness 2.3). If the fingernail scratches some of the stone away, it is likely Gypsum (hardness 2), but if it remains intact (and if a pointed end of it can create a scratch in the fingernail), then it is likely to be Calcium (hardness 3). Common sandpaper at hardness 9 will always scratch Topaz (hardness 8), but will probably not scratch a Ruby or Sapphire (the varieties of Corundum, at hardness 9).

These tools and techniques can be handy in determining hardness relative to one another. However, true hardness (known as "absolute hardness") can only be determined scientifically with the use of a tool called a sclerometer. This gives a much more accurate measure of hardness, with Talc being of absolute hardness 1, Gypsum being 2, Calcite being 9, Fluorite being 21, Apatite being 48, Orthoclase being 72, Quartz being 100, Topaz being 200, Corundum being 400, and Diamond being 1,500.

When a reference to the Mohs scale is given in this book, the reader can turn to this page to get an idea of how hard the particular stone is.

Traditional Birthstones

Official birthstones are first, followed alphabetically by alternate birthstones.

JANUARY: Garnet, Coral, Emerald, Rose Quartz, Sapphire, Serpentine

FEBRUARY: Amethyst, Diamond, Fluorite, Garnet, Kunzite, Onyx, Pearl, Topaz

MARCH: Aquamarine and Bloodstone, Blue Topaz, Hematite, Quartz, Red Jasper

APRIL: Diamond, Coral, Lepidolite, Moonstone, Pearl, Quartz, White Sapphire, Zircon

MAY: Emerald, Agate, Carnelian, Chrysoprase, Garnet, Jade, Tourmaline

JUNE: Pearl, Agate, Alexandrite, Emerald, Moonstone, Sapphire, Turquoise

JULY: Ruby, Carnelian, Diamond, Green Sapphire, Onyx, Pearl, Sardonyx, Turquoise

AUGUST: Peridot, Aventurine, Jade, Sardonyx

SEPTEMBER: Sapphire, Opal, Aquamarine, Coral, Jasper, Tourmaline

OCTOBER: Opal, Jasper, Morganite, Pink Emerald, Pink Sapphire, Pink Tourmaline

NOVEMBER: Topaz, Citrine

DECEMBER: Turquoise, Amazonite, Blue Topaz, Lapis Lazuli, Tanzanite, Zircon

Ayurvedic Birthstones

JANUARY: Garnet

FEBRUARY: Amethyst

MARCH: Bloodstone

APRIL: Diamond

MAY: Agate

JUNE: Pearl

JULY: Ruby

AUGUST: Sapphire

SEPTEMBER: Moonstone

OCTOBER: Opal

NOVEMBER: Topaz

DECEMBER: Ruby

Zodiac Stones

AQUARIUS: Garnet

PISCES: Amethyst

ARIES: Bloodstone

TAURUS: Sapphire

GEMINI: Agate

CANCER: Emerald

LEO: Onyx

VIRGO: Carnelian

LIBRA: Peridot, Aventurine

SCORPIO: Beryl

SAGITTARIUS: Topaz

CAPRICORN: Ruby

Anniversary Stones

Stones are listed alphabetically for each anniversary.

1: Gold, Pearl
2: Garnet, Rose Quartz
3: Pearl, Sapphire
4: Amethyst, Moonstone, Topaz
5: Sapphire
6: Amethyst, Garnet
7: Lapis Lazuli, Onyx
8: Aventurine, Tourmaline
9: Lapis Lazuli, Tiger Eye
10: Black Onyx, Diamond
11: Turquoise
12: Agate, Jade
13: Citrine, Malachite, Moonstone
14: Moss Agate, Opal, Sapphire
15: Ruby
16: Aquamarine, Peridot, Topaz
17: Amethyst, Citrine
18: Garnet, Opal
19: Aquamarine, Topaz
20: Emerald
21: Iolite
22: Spinel
23: Sapphire, Imperial Topaz
24: Tanzanite
25: Silver
30: Diamond, Jade, Pearl, Sapphire
35: Coral, Emerald, Jade
40: Ruby
45: Sapphire
50: Gold, Sapphire
55: Emerald
60: Diamond
65: Star Sapphire
75: Diamond

crystal healing

The use of gemstones in physical healing and health maintenance, and in the promotion of desired attributes and/or banishing of undesirable ones, has been known by humans for thousands of years. Today, modern crystal healers use gems regularly to maintain overall good health, to heal illnesses or injuries, to create positive energy in the body, to protect against negative feelings, to promote skills and positive attributes, and to heal or diminish mental discord. Crystal healing can be done in a variety of ways, and can be done by anyone. While there is no peer-reviewed scientific evidence that shows crystal healing works as intended, it has never been known to do any harm. The one caution a person must take when deciding to undergo crystal healing is that it *not* be a complete substitute for necessary medical treatment, such as prescription drugs or broken bone healing. Used to supplement necessary medical treatment, crystal healing can be a valuable asset.

Following are some general tips to consider when attempting to utilize the energies of gemstones for healing purposes:

To use crystal healing effectively, first choose the stone that corresponds to the situation for which you want help. (See the list on page 122.)

The best way to provide overall healing or maintenance of general health is to wear the stones around your neck. Round beads are thought to carry the best energy, though some people think that the more natural the state of the stone, the more clean its healing ability.

Gemstones can be worn as jewelry; however, their healing properties diminish when they are paired with metals. To get the fullest benefit, stones should be wrapped in or strung on cotton, hemp, silk, or other natural string or cord.

To enhance the effectiveness of a gemstone, crystal healers sometimes pair it with a pure Quartz crystal.

For ongoing aid, a single stone can be worn as a necklace or other jewelry every day. The stone can be rough, tumbled, or cut.

For a good overall optimal health session, lie down and place stones along the body to correspond with the chakras (see page 132). Relax for an hour or two as the stones revive the energy points on the body.

Gemstones can be applied directly to the area needing healing. A bracelet or anklet can be worn to help aid problems with the arms, hands, legs, or feet. Crystals used for heart or stomach health can be worn in a girdle or belt. Stones used to promote love should be worn close to the heart, in a bra or on a string. Stones used for concentration can be worn in a headband or under a hat.

Gemstones can be placed in a room near the user for a steady stream of gentle energy. For example, putting a stone near a student's workstation can enhance the student's ability to concentrate, while putting a different stone near a bed or somewhere in the bedroom can help alleviate insomnia, bad dreams, and other sleep-related problems.

Gemstones can be carried in the pocket and rubbed occasionally for a gentle ongoing stream of healing energy.

Each individual must choose the gemstone that "calls" to him or her. Given the choice of several stones, the user should pick up and hold each one in turn. The "right" stone will have a comfortable feel in the hand.

Gemstones can be used as elixirs by placing the appropriate stone into a jar of pure spring water and letting it soak for up to 24 hours. The water can then be put in an atomizer to be spritzed on the skin and in the room. This water can only be used for a few hours before its healing properties are diminished. Pour the leftover water in the garden or house plants.

Most crystals need to be cleansed after prolonged use. Stones can be cleaned by placing them in moonlight for several nights, by placing them in sunlight, or by placing them in rain. This helps to restore their healing energy.

metaphysical and healing properties

In this section, the healing properties are listed alphabetically, with the stones corresponding to each property listed after them. To use this section, choose the area in which you want information on emotional, physical, spiritual, or mental health, and see which stones are thought to be helpful for that property.

DISCLAIMER: This section is for informational and entertainment purposes only. The reader should not attempt to use crystal therapy to fully replace other healing methods.

ACCEPTANCE: Brecciated Jasper

ADDICTIONS/HABITS: Dumortierite, Iolite, Kunzite, Onyx

ALLERGIES: Brecciated Jasper, Chrysocolla

ANGER: Carnelian, Howlite, Mugglestone, Peridot, Topaz

ANIMAL COMMUNICATION: Brecciated Jasper, Dalmatian Jasper, Zebra Jasper

ANXIETY: Kyanite, Riverstone, Snowflake Obsidian, Variscite, Zebra Jasper

APATHY: Fire Opal

ARTHRITIS: Chrysocolla, Coral, Zebra Jasper

ASTHMA: Serpentine, Topaz

BACK: Fire Opal

BEAUTY: Amber, Red Jasper, Rhyolite, Rose Quartz

BITES/STINGS (POISONOUS): Ametrine, Citrine, Garnet, Red Jasper, Serpentine, Tsavorite

BLEEDING: Bloodstone, Coral, Mugglestone, Topaz

BLOOD: Bloodstone, Brecciated Jasper, Coral, Dalmatian Jasper, Fire Opal, Garnet, Hematite, Mugglestone, Red Jasper, Ruby, Ruby Zoisite, Sapphire

BLOOD PRESSURE: Larvikite, Topaz

BONES: Abalone, Amazonite, Botswana Agate, Brecciated Jasper, Chrysocolla, Coral, Fluorite, Howlite, Riverstone, Tiger Eye, Tourmaline, Turitella, Zebra Jasper

BRAIN: Azurite, Hemimorphite, Labradorite

CALMING: Abalone, Amazonite, Amethyst, Ametrine, Apatite, Aventurine, Bloodstone, Blue Lace Agate, Blue Topaz, Carnelian, Chrysocolla, Crazy Lace Agate, Dumortierite, Goldstone, Hematite, Howlite, Jade, Kambaba Jasper, Kunzite, Kyanite, Labradorite, Larimar, Larvikite, Mookaite, Mugglestone, Nevada Lapis, Prehnite, Red Jasper, Rhodonite, Rose Quartz, Smoky Quartz, Sodalite, Tanzanite, Tourmaline, Variscite

CANCER: Myrrh

CHANCE-TAKING: Bloodstone, Mookaite

CHANGE ACCEPTANCE: Ametrine, Azurite, Botswana Agate, Hessonite, Larvikite, Malachite, Mookaite, Opal, Rhyolite, Riverstone, Sugilite, Tiger Eye, Turitella

CHILDBIRTH: Bloodstone, Malachite, Moonstone, Moss Agate, Myrrh, Picture Jasper, Riverstone

CIRCULATION: Chrome Diopside, Chrysotile, Kunzite, Mugglestone, Red Jasper, Variscite

CLARITY: Hematite, Herkimer Diamond, Kunzite, Kyanite, Lapis Lazuli, Quartz, Rutilated Quartz

COLIC: Malachite, Sapphire

COMFORT: Botswana Agate, Sugilite

COMMUNICATION: Blue Lace Agate, Blue Topaz, Chrysocolla, Peruvian Opal, Pyrite, Rhyolite, Sodalite

CONCENTRATION: Botswana Agate, Kunzite, Kyanite, Pyrite, Quartz, Rutilated Quartz, Variscite, Zebra Jasper

COOLING: Aquamarine, Blue Topaz, Moss Agate

COUGH: Myrrh, Topaz

COURAGE: Amber, Bloodstone, Chrysoprase, Hessonite, Nevada Lapis, Peruvian Opal, Sunstone, Tiger Eye, Tsavorite, Variscite

CREATIVITY: Abalone, Amazonite, Ametrine, Aventurine, Azurite, Botswana Agate, Brecciated Jasper, Carnelian, Chrome Diopside, Fire Opal, Herkimer Diamond, Hessonite, Howlite, Kyanite, Nevada Lapis, Peruvian Opal, Picasso Marble, Rutilated Quartz, Smoky Quartz, Tiger Iron, Tourmaline, Tsavorite

DECISION-MAKING: Abalone, Crazy Lace Agate, Rutilated Quartz, Sodalite, Sunstone

DEPRESSION: Amber, Aquamarine, Botswana Agate, Coral, Dalmatian Jasper, Fire Opal, Kunzite, Lepidolite, Rose Quartz, Sandalwood, Sapphire, Tiger Eye, Topaz, Unakite, Zebra Jasper

DIGESTION: Ametrine, Chrysocolla, Citrine, Kambaba Jasper, Labradorite, Malachite, Moss Agate, Ocean Jasper, Picasso Marble, Sandalwood, Tiger Eye, Turitella

DIZZINESS: Azurite, Peruvian Opal

DREAM ENHANCEMENT: Amethyst, Ametrine, Fluorite, Herkimer Diamond, Jade, Opal, Prehnite

DREAM RECOLLECTION: Brecciated Jasper, Picasso Marble

EARS: Rhodonite, Sapphire

EMOTIONAL HEALTH/BALANCE: Amber, Malachite, Nevada Lapis, Quartz, Rose Quartz, Rutilated Quartz, Tourmaline, Unakite, Zebra Jasper

ENDURANCE: Brecciated Jasper, Chrysocolla, Zebra Jasper

ENERGY AMPLIFICATION: Calcite, Lodalite, Malachite, Prehnite, Quartz, Rhodochrosite, Riverstone, Ruby, Ruby Zoisite, Rutilated Quartz, Tsavorite, Zebra Jasper

ENVIRONMENTALISM/EARTH HEALING: Apatite, Coral, Jet, Mahogany Obsidian, Ocean Jasper, Picture Jasper, Red Jasper, Rhyolite, Smoky Quartz, Turitella, Zebra Jasper

ENVY: Carnelian, Sapphire

EYES: Aventurine, Emerald, Labradorite, Picture Jasper, Tiger Eye, Topaz

FAITH: Pearl, Zebra Jasper

FATIGUE: Peruvian Opal, Turitella, Variscite

FEAR: Brecciated Jasper, Carnelian, Chrome Diopside, Fire Opal, Howlite, Onyx, Rutilated Quartz, Snowflake Obsidian, Sunstone, Turitella, Variscite

FERTILITY: Carnelian, Coral, Ruby Zoisite

FEVER: Aquamarine, Garnet, Lapis Lazuli, Moss Agate, Sandalwood, Sodalite

FORGIVENESS: Rhodochrosite

GENERAL HEALTH: Amber, Ametrine, Citrine, Moss Agate, Peridot, Prehnite, Ruby, Ruby Zoisite

GRIEF: Aquamarine, Chrysocolla, Fire Opal, Jet, Onyx, Smoky Quartz, Snowflake Obsidian

GROUNDING: Carnelian, Dalmatian Jasper, Turquoise

GUILT: Botswana Agate, Chrysocolla, Rutilated Quartz

GUMS: Mahogany Obsidian, Zebra Jasper

HAPPINESS/JOY: Amazonite, Amber, Ametrine, Brecciated Jasper, Cat's Eye, Citrine, Crazy Lace Agate, Dalmatian Jasper, Onyx, Serpentine, Tiger Eye, Topaz, Tourmaline, Unakite, Zebra Jasper

HEADACHE: Amethyst, Ametrine, Azurite, Blue Lace Agate

HEART: Amazonite, Aventurine, Carnelian, Chrome Diopside, Emerald, Garnet, Kunzite, Malachite, Rhodonite, Rose Quartz, Ruby, Ruby Zoisite, Tourmaline

HONESTY: Botswana Agate

HORMONES: Hessonite, Moonstone

HUMOR: Amber, Crazy Lace Agate

IMMUNE SYSTEM: Herkimer Diamond, Hessonite, Larimar, Mookaite, Prehnite, Rhodonite

INFLAMMATION/INFECTION: Moss Agate, Myrrh, Picasso Marble, Sandalwood, Serpentine, Sodalite, Tsavorite

INNOCENCE: Pearl

INSOMNIA: Amethyst, Ametrine, Crazy Lace Agate, Hematite, Howlite, Mugglestone, Peruvian Opal, Sandalwood, Sodalite, Topaz

INTELLECT: Brecciated Jasper, Calcite, Chrome Diopside, Emerald, Fluorite, Hematite, Hessonite, Howlite, Jet, Larvikite, Mugglestone, Nevada Lapis, Pyrite, Ruby Zoisite

INTENSTINES: Fire Opal, Hydrogrossular Garnet

INTUITION: Calcite, Cat's Eye, Moonstone, Tsavorite

INVISIBILITY: Blue Topaz, Chrysoprase, Opal, Topaz

JEALOUSY: Peridot

JOINTS: Abalone

KIDNEYS: Hydrogrossular Garnet, Jasper

KINDNESS: Abalone, Mugglestone, Turquoise

LACTATION: Serpentine

LOVE: Aquamarine, Cherry Quartz, Coral, Emerald, Garnet, Jade, Kunzite, Larimar, Malachite, Moonstone, Rhodochrosite, Rhodonite, Rhyolite, Rose Quartz, Ruby, Topaz, Turquoise

LOYALTY: Dalmatian Jasper, Emerald, Pearl, Malachite

LUCID DREAMING: Jade, Prehnite

LUCK: Abalone, Cat's Eye, Labradorite, Picture Jasper, Turquoise

LUNGS: Aventurine, Chrome Diopside, Emerald, Garnet, Kunzite, Sandalwood

MALE/FEMALE ENERGY BALANCE: Amazonite, Bloodstone, Moss Agate, Onyx, Ruby Zoisite, Tourmaline

MARRIED LOVE: Amazonite, Aquamarine, Emerald, Hydrogrossular Garnet, Onyx, Rose Quartz, Sapphire, Spinel, Tourmaline

MEDITATION AID: Amethyst, Ametrine, Lodalite, Ocean Jasper, Sodalite

MEMORY: Botswana Agate, Calcite, Carnelian, Emerald, Hematite, Howlite, Mugglestone, Pyrite

MENSTRUATION/MENOPAUSE: Moonstone

MENTAL CLARITY: Apatite, Azurite, Dumortierite, Emerald, Kyanite, Labradorite, Larimar, Obsidian, Pyrite, Quartz, Rutilated Quartz, Sunstone

MENTAL STABILITY: Ametrine

MERCY: Aquamarine

METABOLISM: Hessonite, Labradorite, Peruvian Opal, Riverstone

MOUTH: Myrrh

MUSCLES: Aventurine, Abalone, Emerald, Lepidolite, Malachite, Snowflake Obsidian

NEGATIVE ENERGY: Amber, Chrysoprase, Coral, Crazy Lace Agate, Dalmatian Jasper, Fluorite, Kunzite, Onyx, Ruby Zoisite, Snowflake Obsidian, Sodalite, Sugilite, Turquoise

NERVOUS SYSTEM: Amazonite, Kunzite, Variscite

NIGHTMARES: Carnelian, Crazy Lace Agate, Dalmatian Jasper, Lepidolite, Peridot, Snowflake Obsidian, Tsavorite

NUTRITION: Fluorite, Hessonite, Kambaba Jasper, Turitella

OPPOSITE INTEGRATION: Bloodstone

ORDER/ORGANIZATION: Dumortierite

PAIN: Moqui Marbles, Mugglestone, Picture Jasper, Sugilite

PATIENCE: Dalmatian Jasper, Howlite, Moss Agate, Ocean Jasper, Rhodonite

PEACE: Blue Lace Agate, Jade, Kambaba Jasper, Kunzite, Larimar, Moonstone, Moss Agate, Nevada Lapis, Ocean Jasper, Rhodonite, Sapphire

POSITIVITY: Amber, Crazy Lace Agate, Fluorite, Peridot, Ruby Zoisite, Snowflake Obsidian

POWER: Ametrine, Citrine, Jet

PREGNANCY: Moss Agate, Unakite

PROBLEM-SOLVING: Botswana Agate, Chrysoprase, Obsidian, Sodalite, Tiger iron

PROSPERITY: Amazonite, Amethyst, Ametrine, Aventurine, Calcite, Cat's Eye, Citrine, Coral, Jet, Onyx, Ruby, Ruby Zoisite, Snowflake Obsidian, Tiger Eye

PROTECTION: Botswana Agate, Coral, Moonstone, Mugglestone, Nevada Lapis, Peridot, Red Jasper, Tiger Eye, Tourmaline

RADIATION: Hematite, Malachite, Sodalite

RELATIONSHIPS: Chrome Diopside, Chrysocolla, Chrysoprase, Dalmatian Jasper, Dumortierite, Hydrogrossular Garnet, Iolite, Kambaba Jasper, Malachite, Mugglestone, Picasso Marble, Picture Jasper, Rhodochrosite, Rhodonite, Rose Quartz, Ruby, Ruby Zoisite, Sapphire, Smoky Quartz, Sodalite, Turquoise, Turitella

RELAXATION: Amethyst, Ametrine, Dalmatian Jasper, Moqui Marbles, Sodalite, Zebra Jasper

REPRESSION: Fire Opal, Rutilated Quartz, Unakite

REPRODUCTIVE ORGANS: Carnelian, Chrysoprase, Coral (female), Moss Agate (female), Moonstone (female), Ruby Zoisite, Unakite, Variscite (male)

RESPECT: Hessonite, Picasso Marble, Serpentine

SELF-CONFIDENCE: Amber, Ametrine, Azurite, Chrysoprase, Citrine, Jade, Kunzite, Labradorite, Peruvian Opal, Quartz, Rhodochrosite, Rhodonite, Rose Quartz, Ruby Zoisite, Rutilated Quartz, Sodalite, Sugilite, Zebra Jasper

SELF-DISCIPLINE: Ametrine, Citrine, Kunzite, Picasso Marble

SELF-EMPLOYMENT/LEADERSHIP: Bloodstone, Sunstone

SELF-RESPECT: Hessonite, Kunzite, Picasso Marble, Quartz, Rhodochrosite, Rhodonite, Rutilated Quartz

SEXUALITY: Bloodstone, Botswana Agate, Cherry Quartz, Mahogany Obsidian, Rose Quartz, Sandalwood

SKIN: Chrysotile, Coral, Crazy Lace Agate, Hemimorphite, Myrrh, Rose Quartz, Sandalwood, Sapphire, Tsavorite

SLEEP DISORDERS: Crazy Lace Agate, Fluorite, Red Jasper, Sandalwood, Tsavorite

SOBRIETY: Amethyst, Ametrine

SPINE: Emerald, Garnet

SPIRITUAL HEALTH: Amber, Hessonite, Jade

STAMINA: Botswana Agate, Crazy Lace Agate, Zebra Jasper

STRENGTH: Chrysocolla, Fire Opal, Mahogany Obsidian, Tiger Eye, Tsavorite, Zebra Jasper

STRESS: Amazonite, Ametrine, Apatite, Chrysocolla, Fire Opal, Hemimorphite, Howlite, Kambaba Jasper, Labradorite, Lepidolite, Moss Agate, Mugglestone, Onyx, Peruvian Opal, Sandalwood, Sunstone, Variscite

STUBBORNESS: Dumortierite

SUCCESS: Ametrine, Botswana Agate, Citrine, Serpentine, Topaz, Zebra Jasper

SUNBURN: Aquamarine, Chrysotile

TEETH: Amazonite, Coral, Fluorite, Howlite, Riverstone, Turitella, Zebra Jasper

TENSION RELEASE: Ametrine, Variscite

THROAT: Amazonite, Aquamarine, Blue Lace Agate, Lapis Lazuli, Myrrh

TOLERANCE: Rhodochrosite

TOOTHACHE: Amethyst, Ametrine, Aquamarine, Zebra Jasper

TOXIN REMOVAL: Ametrine, Azurite, Botswana Agate, Herkimer Diamond, Iolite, Kambaba Jasper, Ocean Jasper, Snowflake Obsidian, Tiger Iron, Turquoise

TRUST: Botswana Agate, Chrome Diopside, Pearl, Sapphire, Sodalite, Turquoise

ULCERS: Chrysocolla, Tiger Eye

VOICE: Blue Lace Agate

WEIGHT GAIN: Unakite

WEIGHT LOSS: Apatite, Peruvian Opal, Picasso Marble, Riverstone, Rose Quartz

WISDOM: Blue Lace Agate, Chrysocolla, Hessonite, Jade, Jet, Kambaba Jasper, Lapis Lazuli, Sapphire, Serpentine, Tsavorite, Turquoise

WOUNDS: Chrysoprase, Kambaba Jasper

chakras

PLEASE NOTE: This is only a very rudimentary summary of chakra healing. Those wanting to use gemstones to heal the body by aligning the chakras should get more in-depth information on how this form of medicine works.

In traditional Indian medicine, the chakras (from a Sanskrit word meaning "wheel" or "turning") are "force centers," or wheels of energy coming from different areas of the body. There are seven basic chakras (in addition to hundreds of secondary chakras), aligned along the spine from the top of the head down to the bottom of the abdomen. When they are properly aligned (such as the cogs of a well-working clock), the body experiences optimal physical, mental, and spiritual health.

When one or more of the chakras becomes misaligned (think of it as a single cog in a clock falling out of line), the body can experience illness or mental discord. The diagnosis of an illness tells the healer which chakra is misaligned. Realigning this particular chakra is said to remove the illness. Then, steps are taken to cleanse and clear the unbalanced or blocked chakra, to get the energy flowing through the body freely again.

Each chakra is associated with certain colors, sounds, and scents. Realignment of the chakras includes treatment with the tones associated with the affected chakra, fragrances that are associated with it, and objects of colors that correspond with the affected chakra's governing color. Gemstones of specific colors are used frequently in this type of healing.

The seven basic chakras, their locations, their associated functions, and their colors are very briefly outlined as follows:

1. **Muladhara**: Base Chakra or Root Chakra. Located at the very base of the spine, near the ovaries or prostate. Representing Earth, survival, stillness. The grounding chakra; it grounds us in the physical world. Promotes acceptance. Problems are manifested as resentment or rigidity. Color: RED.

2. **Swadhisthana**: Sacral Chakra. Located at the last bone of the spine (the coccyx), just beneath the level of the navel. Related to sexuality and reproduction. Promotes creativity. Problems are manifested as emotional imbalance or sexual guilt. Color: ORANGE.

3. **Manipura**: Solar Plexus Chakra. Located in the navel area. This chakra is the seat of emotions. Related to commitment. Problems are manifested as anger or greed. Color: YELLOW.

4. **Anahata**: Heart Chakra. Located in the heart area. The seat of the soul in the hourglass of time. Related to compassion. Problems are manifested as fear or insecurity. Color: GREEN.

5. **Vishuddha**: Throat Chakra. Located in the throat/neck area. Tied to creativity, communication, and truth. Problems are manifested as denial or rudeness. Color: LIGHT BLUE.

6. **Ajna**: Brow Chakra or Third Eye Chakra. Located in the third eye area, at the base of the brain. Governs all mental activities, including dreaming, clairvoyance, intuition, and imagination. Problems are manifested as confusion or depression. Color: INDIGO.

7. **Sahasrara**: Crown Chakra. Located at the top of the head. It governs spirituality, knowledge, and relationship with God. Problems are manifested as grief. Color: VIOLET.

When using gemstones to align chakras, the general rule is that the color of the gemstone should correspond with the color represented by the particular chakra being treated. For example, most yellow-colored stones can be used for the Solar Plexus Chakra. However, there are some exceptions to this. Many stones work well with several different chakras. Often one stone can cover two or three adjacent chakras. It is important to remember that this is not an exact science, and sometimes a certain stone typically used for one chakra might work better with an adjacent chakra, depending on the individual person and/or the color of the particular stone in use. On the next page is a list of the seven chakras, and the stones found in this book that can be used to correspond to each chakra.

chakra stones

ROOT/BASE CHAKRA
Abalone, Bloodstone, Botswana Agate, Brecciated Jasper, Chrysotile, Coral (red), Dalmatian Jasper, Garnet, Hematite, Howlite, Hydrogrossular Garnet, Jet, Kambaba Jasper, Labradorite, Larvikite, Mahogany Obsidian, Mookaite, Mugglestone, Myrrh, Obsidian, Onyx, Opal, Picasso Marble, Pyrite, Quartz, Red Jasper, Ruby, Rutilated Quartz, Smoky Quartz, Snowflake Obsidian, Spinel, Tourmaline, Turitella, Zebra Jasper.

SACRAL CHAKRA
Abalone, Amber, Carnelian, Fire Opal, Garnet, Goldstone, Hessonite, Howlite, Mugglestone, Opal, Quartz, Sandalwood, Sunstone, Tourmaline.

SOLAR PLEXUS CHAKRA
Abalone, Amber, Apatite, Citrine, Coral (black), Fire Opal, Hessonite, Howlite, Jade (yellow), Mookaite, Ocean Jasper, Opal, Pearl, Picture Jasper, Pyrite, Quartz, Rutilated Quartz, Tiger Eye, Tourmaline.

HEART CHAKRA
Abalone, Amazonite, Aventurine, Bloodstone, Calcite, Chrome Diopside, Chrysoprase, Coral (pink), Emerald, Fluorite, Howlite, Jade (green), Kunzite (pink), Lepidolite, Malachite, Moss Agate, Muscovite, Nevada Lapis, Ocean Jasper, Opal, Peridot, Prehnite, Pyrite, Quartz, Rhodochrosite, Rhodonite, Rhyolite, Rose Quartz, Tourmaline, Tsavorite, Unakite, Variscite.

THROAT CHAKRA
Abalone, Amazonite, Amber, Apatite, Aquamarine, Azurite, Blue Lace Agate, Blue Topaz, Chrysocolla, Crazy Lace Agate, Dumortierite, Fluorite, Goldstone (blue), Hemimorphite, Howlite, Kyanite, Lapis Lazuli, Larimar, Lepidolite, Ocean Jasper, Opal, Peruvian Opal, Quartz, Sapphire, Sodalite, Tanzanite, Tourmaline, Turquoise.

BROW/THIRD EYE CHAKRA
Abalone, Ametrine, Azurite, Fluorite, Howlite, Iolite, Kunzite (purple), Kyanite, Lapis Lazuli, Larimar, Lepidolite, Mookaite, Moonstone, Moqui Marble, Opal, Opalite, Quartz, Rainbow Calsilica, Sodalite, Tanzanite, Tourmaline.

CROWN CHAKRA
Abalone, Amethyst, Azurite, Botswana Agate, Brecciated Jasper, Diamond, Herkimer Diamond, Howlite, Larimar, Lepidolite, Lodolite, Moonstone, Moqui Marble, Myrrh, Ruby Zoisite, Rutilated Quartz, Serpentine, Sugilite, Tanzanite, Tourmaline.

Index

This index contains all of the stones listed in this book, as well as alternate names for the stones. When you cannot find a stone in the main text, you might find it in the index.

Abalone, 3
African Jade, 47
Agate, 4
Amazonite, 5
Amazonstone, 5
Amber, 6-7
Amethyst, 8
Amethyst-Citrine Quartz, 9
Ametrine, 9
Antigorite, 94
Anyolite, 90
Apache Tears, 60, 70
Apatite, 10
Appaloosa Jasper, 111
Appaloosa Magnesite, 111
Aquamarine, 11
Asparagus Stone, 10
Atlantis Stone, 71
Australian Jade, 23
Australian Rainforest Jasper, 86
Aventurine, 12
Aventurine Glass, 41
Aventurine Orthoclase, 101
Azurite, 13
Balas Ruby, 99
Banded Chalcedony, 15
Beryl, 11, 33
Black Amber, 50
Black Moonstone, 54, 57
Blood Jasper, 14
Bloodstone, 14
Blue Andean Opal, 76
Blueberry Quartz, 38
Bluebird, 13

Blue Lace Agate, 15
Blue Pearl, 57
Bluestone, 98
Blue Topaz, 16, 104
Botswana Agate, 17
Brecciated Jasper, 18
Cadmia, 43
Caeruleum, 13
Cairngoran, 96
Calamine, 43
Calcite, 19
Canadian Blue Stone, 98
Cape Emerald, 79
Carnelian, 20
Cat's Eye, 35, 103
Cat's Eye Apatite, 10
Champagne Garnet, 45
Chandan, 92
Cherry Opal, 36
Cherry Quartz, 38
Chlorophane, 34
Chrome Diopside, 21
Chromium Diopside, 21
Chrysocolla, 22
Chrysolite, 75
Chrysoprase, 23
Chrysotile, 24
Cinnabar, 25
Cinnamon Garnet, 45
Citrine, 26
Coon Tail, 96
Copal, 6-7
Coral, 27
Cordierite, 48

Cornelian, 20
Corundum, 89, 93
Crazy Horse Jasper, 111
Crazy Horse Magnesite, 111
Crazy Lace Agate, 28
Crocidolite, 103
Crocodile Jasper, 51
Cupid's Darts, 91
Dalmatian Jasper, 29
Dalmatian Rock, 29
Dalmatian Stone, 29
Dalmatine, 29
Dalmatine Jasper, 29
Diamond, 30-31
Dichroite, 48
Disthene, 53
Dogtooth Amethyst, 8
Dragon's Eye, 103
Dumortierite, 32
Dumortierite Quartz, 32
Elbaite, 105
Emerald, 33
Epidote, 109
Epidotized Granite, 109
Evening Emerald, 75
Eye Agate, 17
Fiber Optic Cat's Eye, 35
Fire Opal, 36
Flowering Obsidian, 97
Fluorite, 34
Fool's Gold, 80
Fossil Agate, 108
Fossil Limestone, 108
Fossil Stone, 108
Freshwater Pearl, 37
Fruit Quartz, 38
Galmei, 43
Garden Quartz, 59
Garnet, 39
Gem Silica, 22
Gerisol, 74
Gold, 40

Golden Amethyst, 9
Goldstone, 12, 41
Gooseberry Stone, 47
Grape Jade, 79
Hackmanite, 98
Hawk's Eye, 103
Heliolite, 101
Heliotrope, 14
Hemalyke, 42
Hematine, 42
Hematite, 43
Hemimorphite, 43
Herkimer Diamond, 44
Hessonite, 45
Hiddenite, 52
Howlite, 46
Hydrogrossular Garnet, 47
Iceland Spar, 19
Imperial Jade, 49
Imperial Red Jasper, 83
Imperial Topaz, 104
Inca Rose, 84
Inclusion Quartz, 59
Indian Jade, 12
Indicolite, 105
Infinite Stone, 94
Iolite, 48
Iron Pyrite, 80
Ironstone Concretion, 64
Jade, 49
Jadeite, 49
Jagate, 4
Jasp-Agate, 4
Jasper, 4
Jet, 50
Kambaba Jasper, 51
Kambaba Stone, 51
Kunzite, 52
Kyanite, 53
Labradorite, 54
Lace Agate, 28
Lal, 99

Landscape Jasper, 78
Landscape Quartz, 59
Lapis Lazuli, 55
Lapis Nevada, 69
Larimar, 56
Larvikite, 57
Lavenderine, 58
Lemon Quartz, 38
Leopardskin Jasper, 86
Lepidolite, 58
Lignite, 50
Lilalite, 58
Limestone, 19
Lime Quartz, 38
Lodalite, 59
Lodolite, 59
London Blue Topaz, 16
Lunar Quartz, 74
Luvalite, 100
Madeira Citrine, 26
Mahogany Obsidian, 60, 70
Malachite, 61
Marble, 19
Marcasite, 80
Mexican Agate, 28
Mexican Fire Opal, 36
Mexican Lace Agate, 28
Mica, 67
Mocha Agate, 65
Mocha Stone, 65
Mochi Marble, 64
Monk's Gold, 41
Monkstone, 41
Mookaite, 62
Mookite, 62
Mook Jasper, 62
Moon Jewel Jasper, 71
Moonstone, 63
Moonstone Quartz, 74
Moqui Marble, 64
Morion, 96
Moroxite, 10

Moss Agate, 65
Mother of Pearl, 37
Moukaite, 62
Mugglestone, 66
Muscovite, 67
Muscovite Quartzite, 67
Myrrh, 68
Mystic Topaz, 104
Navajo Cherry, 62
Nephrite, 49
Nevada Lapis, 69
Nevada Stone, 69
New Jade, 94
Norwegian Moonstone, 57
Norwegian Pearl Granite, 57
Obsidian, 70
Obsidian Bomb, 70
Ocean Jasper, 71
Olivine, 75
Onyx, 72
Opal, 73
Opalite, 74
Opalized Quartz, 74
Optical Calcite, 19
Orange Topaz, 104
Orbicular Jasper, 71, 86
Ox Eye, 103
Padparadscha, 93
Patuxent River Agate, 87
Patuxent River Stone, 87
Paua Shell, 3
Peace Jasper, 69
Peace Stone, 69
Peacock Opal, 76
Pearl, 37
Pearlspar, 57
Pectolite, 56
Peridot, 75
Perthite, 5
Peruvian Opal, 76
Picasso Jasper, 77
Picasso Marble, 77

Picasso Stone, 77
Picture Jasper, 78
Picture Rock, 78
Pietersite, 103
Pigeon Blood Agate, 20
Pineapple Quartz, 38
Pink Quartz, 88
Pink Turquoise, 83
Poppy Jasper, 18
Prehnite, 79
Pumice, 86
Purple Turquoise, 100
Pyrite, 80
Quartz, 81
Quartzite, 67
Rainbow Calsilica, 82
Rainbow Obsidian, 70
Rainforest Jasper, 86
Raspberry Spar, 84
Red Jasper, 83
Red Labradorite, 101
Red Sand Stone, 41
Rhodochrosite, 84
Rhodonite, 85
Rhyolite, 86
Ribbon Jasper, 78
Riverstone, 87
Rock Crystal, 81
Rosa del Inca, 84
Rose de France, 8
Rose Quartz, 88
Rose Topaz, 104
Royal Azel, 100
Rubellite, 105
Ruby, 89
Ruby Zoisite, 90
Russian Diopside, 21
Rutilated Quartz, 91
Rutile Quartz, 91
Ryolite, 86
Sagenite, 91
Sandalwood, 92

Sand Stone, 41
Sapphire, 93
Sard, 20
Sardonyx, 20, 72
Scenic Jasper, 78
Scenic Quartz, 59
Schorl, 105
Sea Opal, 3, 74
Serpentine, 94
Shaman Stone, 64
Sherry Topaz, 104
Silver, 95
Sky Blue Topaz, 16
Smithsonite, 43
Smokey Quartz, 96
Smoky Quartz, 96
Smoky Topaz, 96
Snake Agate, 71
Snowflake Obsidian, 70, 97
Snow Quartz, 81
Sodalite, 98
Spectrolite, 54
Spherulitic Obsidian, 97
Spinel, 99
Spodumene, 52
Stellaria, 41
Strawberry Quartz, 38
Stromatolite, 51
Sugilite, 100
Sun Sitara, 41
Sunstone, 101
Swiss Blue Topaz, 16
Tanzanite, 102
Thulite-Diopside Skarn, 69
Tiger Eye, 103
Tiger Iron, 66
Topaz, 104
Tourmalated Quartz, 105
Tourmalinated Quartz, 105
Tourmaline, 105
Transvaal Jade, 47
Tree Agate, 65

Trystine, 9
Tsavolite, 106
Tsavorite, 106
Turitella, 108
Turquanite, 46
Turquoise, 107
Turquonite, 46
Turritella, 108
Turritella Agate, 108
Ulexite, 35
Unakite, 109
Utahlite, 110
Variscite, 110
Venus' Hair Stone, 91
Verdilite, 105
Vertilite, 21
Watermelon Quartz, 38
Watermelon Tourmaline, 103
Water Sapphire, 48
White Asbestos, 24
White Buffalo Stone, 46
White Turquoise, 111
Wild Horse Jasper, 111
Wild Horse Magnesite, 111
Wild Horse Picture Jasper, 111
Wild Horse Turquoise, 111
Windalia Radiolarite, 62
Zebra Agate, 112
Zebra Jasper, 112
Zebra Marble, 112
Zebra Rock, 112
Zebra Stone, 112
Zoisite, 90

about the author

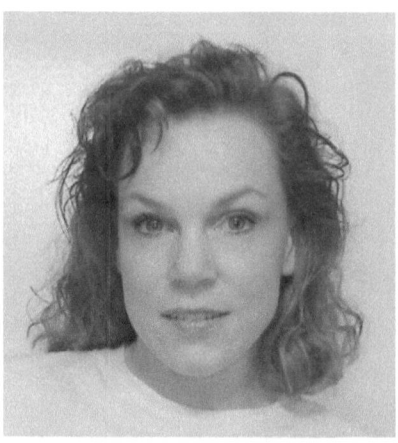

Suzanne Bettonville first got interested in gems as a teenager. Growing up on a farm in Oregon, she found lots of beautiful samples of rose, clear, and white quartz in her yard, and always hungered to learn more about them. She was a member of the Rogue Gem & Geology Club in southern Oregon as a teenager, attending rock hunting expeditions and manning county fair booths for the club.

Suzanne went to Southern Illinois University at Edwardsville and studied Anthropology, with Earth Science as a minor field. Her studies in Earth Science fulfilled her earlier fascination with gemstones, and created an ongoing quest for information about minerals and gems. She was recruited to do scientific illustrations for one of her Earth Science professors, and maintained a love of rocks and minerals into adulthood.

Suzanne owns and operates the online store Ephemerala Natural Stone Jewelry & Gifts (website at www.ephemerala.com), which specializes in gems and gemstone products as well as native art, incense, and musical instruments from around the world.

Suzanne currently lives in southern Utah with four sons. She is active in local theater, does research into her family history, and engages in all manner of artistic and creative pursuits.

www.ingramcontent.com/pod-product-compliance
Lightning Source LLC
Chambersburg PA
CBHW021943170526
45157CB00003B/902